AF709172

# Telephone Projects

## For Electronics Experimenters

Carl J. Bergquist

# Telephone Projects

## For Electronics Experimenters

Carl J. Bergquist

PROMPT®
PUBLICATIONS

©1998 by Howard W. Sams & Company

**PROMPT© Publications** is an imprint of Howard W. Sams & Company, A Bell Atlantic Company, 2647 Waterfront Parkway, E. Dr., Indianapolis, IN 46214-2041.

All rights reserved. No part of this book shall be reproduced, stored in a retrieval system, or transmitted by any means, electronic, mechanical, photocopying, recording, or otherwise, without written permission from the publisher. No patent liability is assumed with respect to the use of the information contained herein. While every precaution has been taken in the preparation of this book, the author, the publisher or seller assumes no responsibility for errors or omissions. Neither is any liability assumed for damages resulting from the use of information contained herein.

**International Standard Book Number: 0-7906-1168-6**
**Library of Congress Catalog Card Number: 98-067314**

***Acquisitions Editor***: Loretta Yates
***Editor***: Pat Brady
***Assistant Editors***: Natalie Harris, J.B. Hall
***Typesetting***: Pat Brady
***Cover Design***: Christy Pierce
***Graphics Conversion***: Terry Varvel
***Illustrations and Other Materials***: Courtesy of the Author

***Trademark Acknowledgments***:
All product illustrations, product names and logos are trademarks of their respective manufacturers. All terms in this book that are known or suspected to be trademarks or services have been appropriately capitalized. PROMPT© Publications, Howard W. Sams & Company, and Bell Atlantic cannot attest to the accuracy of this information. Use of an illustration, term or logo in this book should not be regarded as affecting the validity of any trademark or service mark.

PRINTED IN THE UNITED STATES OF AMERICA

9 8 7 6 5 4 3 2 1

# Contents

# INTRODUCTION

Welcome to the wonderful world of telephones!...and, of course, related subjects and equipment. *Telephony*, as it is sometimes called, is a remarkable technology that has grown to be indispensable in our culture. Today, where would we be without telephones, and the telephone system?

Aside from being able to conduct business conversations, and talk with friends and family, this system allows us to communicate in a variety of ways. Modems permit our personal computers to interact with other computers virtually all over the planet. Additionally, telephone lines represent an “electronic interstate” system that affords a multitude of control functions. But all this will be covered in more detail as we go along.

For most of its life, the telephone system existed as a deep dark secret, not to be bridged by intruders. It reminded me of the Winston Churchill quote (I think it was him) regarding the Soviet Union being, “an enigma, surrounded by paradox, wrapped in a mystery,” or something like that. And, that was just how the telephone company (telco) wanted it.

However, in the later 1970s, “Ma Bell” had to open her system to the outside world, and this forever changed the picture. Now, equipment made by manufacturers other than the phone company could be placed on the lines, as long as said equipment didn’t cause any interference or damage.

That, of course, gave rise to a whole new industry, as well as generating a fresh arena for us hobbyists. The commercial market was soon flooded with an assortment of novel devices, and in this text we will discuss some of those. We will also build some interesting and useful stuff.

I have included two telephones, a modem, a dual-tone multifrequency (DTMF) controller, a tape recorder switch and more. But best of all is the Telephone Lab. With this unit you will have the tools necessary to test telephone equipment, and design your own special devices. This one I think you will like!

So, if you have an interest in telephones and associated gear, this book is for you. The text will furnish an understanding of the fundamentals, some "hands-on" projects, and a short history of how the telephone came to be.

I would, however, be remiss in this introduction if I didn't confess to a lingering fascination with the telephone system. In short, the mystique lives on. Every time I start to plug an RJ11 into a modular jack, I have this mental image of a demon reaching out and snatching me in!...that I will forever ride the telephone lines, in limbo, like some sort of electron (Ooooo-Eeee-Ooooo!).

Nah! Not really! Actually, there isn't a whole lot to fear from the telephone system; it's fairly harmless. However, for reasons that will become perfectly clear in the next chapter, I wouldn't recommend holding onto the ring and tip (red and green) leads when the phone rings!

Anyway, have some fun! Tackle those demons (if they exist)! For, with the knowledge this book provides, you will find an enchanting world on those lines — one that will serve you well! The versatility of the phone system is rivaled by few others. So, go ahead! Put it to work for you!

# CHAPTER 1

# TELEPHONE SYSTEM OVERVIEW

## A TELEPHONE CALL

In this chapter, we will look at how the telephone system functions. In reality, it isn't as complicated as certain parties would like you to believe. The various operations managed by the system are governed primarily by voltage potentials, and changes in those potentials.

In the way of an illustration, let's say Dick is going to call Jane. No, no....uh, I think those two have retired following a long and exhausting career educating first-graders. Instead, we will say that Loretta is going to call Natalie. (This is in tribute to the managing editor and assistant editor of this book—old friends of mine. Yeah, I know....some folks would call it something else, but that's their problem.)

Before Loretta picks up the telephone receiver to make the call, her phone is in the "on-hook" status. This means that a direct current (DC) voltage of between 36 and 50 volts is present across the "tip" (green wire) and "ring" (red wire) of the telephone line. This is known as the "battery voltage." Nominally, it will be about 48 volts, but that varies among telephone companies (telcos).

As a quick sidebar, the terms "tip" and "ring" come from the early days of telephone service. That was back when you had a real live operator, at a switchboard, engaging and disengaging "phone" style plugs to connect one telephone to another. *Tip* referred to the "knob shaped" end of those plugs, while *ring* alluded to the metal shaft, or outer ring. Today, these designations have endured, probably for "old times" sake.

Now, when Loretta lifts the handset, the phone puts a load on the line, and the 48 volts drops to between 2 and 15 volts. Again, normally this will be about 5 to 7 volts DC. Since her local handy-dandy telco central switching office (CO) is monitoring her line, as well as everyone else's, it detects this voltage drop, and impresses a dial tone on Loretta's line.

The dial tone is CO-generated, and is a tone pair of 350 hertz and 440 hertz. It is also a "passive" action, as it is present only to let Loretta know that her phone line is working.

Loretta now dials Natalie's number using either a rotary "pulse" dial, or pressing numbered buttons that produce "dual-tone multifrequency" (DTMF) tone pairs. I will get into the dialing schemes in more detail later, but for now, it is only necessary to know that Loretta has communicated to the CO the number of the phone she wishes to reach.

The central office, through switches and computers, then checks Natalie's line. If it sees 48 volts, it knows Natalie's phone is on-hook, and two things happen: First, it rides a 20-hertz alternating current (AC) "ringer voltage" of between 90 and 110 volts on Natalie's line. This activates the ringer circuit in her phone; Second, it impresses a CO-generated "ringer signal" on Loretta's line. This again is a passive operation to let Loretta know the central office is trying to connect her to Natalie.

By the way: Remember in the Introduction when I said it would become apparent in this chapter why you don't want to be touching the ring and tip lines whenever the phone rings? Well, that 90 to 110 volts AC is the reason. Even though the current is low, it would feel a lot like sticking your fingers into an empty, and HOT, lamp socket.

OK! So much for that. If, however, the central office sees a 5-volt potential on Natalie's line, then it knows her phone is "off-hook." First, the CO will check to see if Natalie has "call waiting," and if she does, Loretta will receive the normal ringer tone, while Natalie will hear a periodic "beep" to indicate another call is waiting for her.

If Natalie doesn't have the call waiting service, then Loretta will hear the familiar "busy signal." This, of course, is another CO-generated tone, and tells Loretta to hang up and try later.

One last passive signal to be considered is the "off-hook warning" tone. I would imagine everyone is acquainted with that annoying "beep...beep...beep" noise that lets you know the phone is off-hook. It is initiated after 10 to 15 seconds if the line voltage drops from 48 to 5 volts, and no dialing pulses or tones are detected at the central office.

This warning is sometimes preceded by a couple of ringer tones, then an announcement. That announcement is usually to the effect that "if you want to make a call, please hang up and try again." Maybe this an effort on the part of some telcos to be polite. It's a nice thought, anyway.

Thus, as previously stated, functions of a typical "home telephone" system are controlled by changes in the line-level voltage. And that makes for a rather simple, but highly efficient, process — for the most part, one that has not been radically altered since the inception of telephone service.

Through this configuration, Loretta can call Natalie, or Pat, or Kelli, or Terry, or me whenever she wants to. And, with relative ease and speed. Naturally, there is a little more to the "big picture" than has been presented here, but for our purposes, the preceding provides an adequate explanation of a typical "end-to-end" (telephone to telephone) call.

## THE CENTRAL SWITCHING OFFICE (CO)

With all that said, let's take a closer look at what the telephone company lovingly refers to as a "central switching office." First off, if you live in an area with 100,000 or more telephones, there is likely to be more than one switching office. From what I gather, there isn't a specific set of rules governing the number of COs, or exchanges (the first three digits of a telephone number) assigned to each switching office. This all appears to be up to the discretion and capability of each individual telco.

For example, in a city like mine where 70 different exchanges are used, there are three central offices and one "non-switching" wire center, the latter being simply a wire connection, or "terminal" facility.

As for the exchange (prefix) breakdown, one CO handles 26, a second manages 38 and the last controls only 6. Of course, this last CO will be expanded as service needs increase. Incidentally, the number of exchanges will give you a "ballpark" figure of how many telephones are in use (within 9,998). This is because each exchange can handle 9,999 numbers (for example, 271-0000 to 271-9999). So, by multiplying the total number of exchanges by 9,999, you will have the maximum number of phones the system can manage. Subtracting 9,998 will reveal the minimum number of telephones.

Knowing this, I can estimate the number of phones in my area to be between 689,932 and 699,930. Naturally, this only covers the conventional "wired" phones, or "land lines," and not the wireless, or "cellular" units.

OK, getting back to Loretta calling Natalie. If both phones are in the same exchange, let's say "384," then the call will be handled by a single CO. However, let's say Natalie's exchange, "278," comes under the jurisdiction of another CO. Now the central office that handles 384 will have to hand the call off to the CO that covers 278.

In this fashion, all information regarding the call, as well as the passive tones, will be communicated back and forth between the two offices. This does add a degree of complexity to the system, but to date it has all worked quite well.

In the event Loretta wants to place a "long-distance" call, her local CO will first detect the call, then hand it off to an "interexchange office" (IO). The IO will transmit the call to a second IO at the destination, and that signal will be routed to the central office that handles whoever Loretta is phoning.

As might be expected, this is where "area codes" come into play. By adding the three additional digits to the dialed number, the phone company knows where the call is going, and routes it accordingly. Also, adding the digit "1" to the beginning of the dial sequence informs the central office a long-distance call is being placed.

Briefly addressing the methods used to move calls from one phone to another, there are three primary vehicles employed: (1) cable, or physical wires between the two; (2) microwave, which passes the call from one substation to the next via super high-frequency (SHF) radio waves; and (3) satellites, which receive and relay calls from earth while sitting in a "geostationary" orbit above the planet.

In the last system, SHF radio waves are again employed, and transcontinental communication is possible. Before satellites, calls from North America to Europe, or elsewhere, had to go by undersea coaxial cable, or shortwave radio.

All three systems are readily used today, with direct cable connections most common within local telephone systems. However, as technology advances, both microwave and satellite transmissions may well become customary, even in local service.

One last subject that deserves mention is "fiber optic," or "lightwave" communications. Using miniature semiconductor lasers and very fine glass fibers, exceptional quality can be achieved.

Sound from the telephone is encoded and used to "modulate" the light beam. It is then transferred over the fiber optics and decoded at the opposite end. While this technique is still in its infancy, such viable factors as cost effectiveness and reliability are definitely moving fiber optic systems forward.

## TYPES OF TELEPHONES AND TELEPHONE SYSTEMS

Virtually all telephones have certain elements in common. All require both a transmitting device (microphone) and receiver (speaker/headphone). Also, all have to have some form of "number dialing" scheme. The latter is usually either a rotary pulse dial, or a pushbutton "touch-tone" DTMF keypad.

Regarding the rotary dials we all used for so many years (well, some of us anyway), these are based on the principle of sending individual pulses, or "clicks," to the central office. When you dial an "8," eight clicks are "heard" at the CO. For a "2," two clicks, and so forth.

This system served telephony well for many many years, but it was slow, and somewhat error prone. Thus, after World War II, new methods were explored. This was as much an effort to keep up with the ever-increasing demand for telephone service as anything else. Much more about that, and related subjects, is found in Chapter 11.

After a lot of research, the dual-tone multifrequency DTMF concept was adopted to eventually replace rotary dialing. In this strategy, each digit, as well as the pound (#) and asterisk (*) keys, are represented by dual tones or a "tone pair." Each tone pair consists of one high and one low audio frequency.

You may have noticed that everything sent over conventional phone lines is in the audio frequency range (roughly 20 to 20,000 hertz). This is because of the very nature of its original intended application (sending voice over a wire). Hence, the phone system was designed for those frequencies. Since that's still the primary function, hey! — you know, if it ain't broke, don't fix it.

Anyway, back to the DTMF tones. Each time a keypad button is pressed, two distinct audio tones are generated that tell the CO which key has been activated. For example, the digit "2" key will produce 697 hertz and 1336 hertz (see *Figure 1-1*). Both are placed on the line simultaneously, and one is from the "high" group, while the other is from the "low" group. You take the high road, and I'll take the.....never mind.

Why two tones you ask? Well, that's a reasonable question. The answer is that it's done to prevent what phone company people like to call "talk-off." What is talk-off? Another reasonable question. Talk-off is a dialing error that occurs when the audio signal at the transmitter (voice, music, etc.) imitates the dialing tones, and falsely dials a digit.

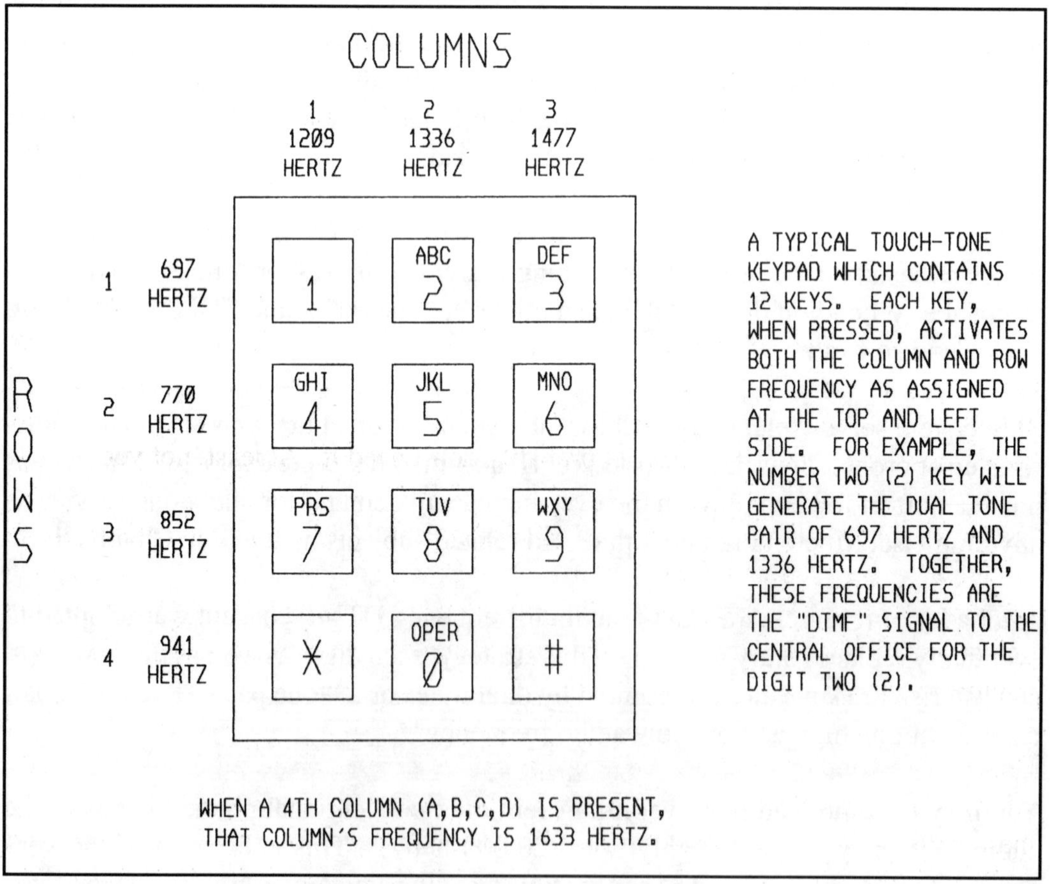


**Figure 1-1.** DTMF frequencies.

If each number was represented by a single tone, talk-off would be a major dilemma, as imitated tones would be common. However, by having two tones, it becomes substantially more difficult to imitate both of them at the same time, especially when those audio frequencies are reasonably separated.

In reality, there are some additional precautions installed to help prevent talk-off. These usually come in the form of filters, as it is possible for some music to produce a wide range of frequencies. However, such techniques get complicated, and it's really only necessary to be aware of their existence.

As for mobile, or "cellular" phones, to my knowledge all such devices utilize DTMF pushbutton dials. I say "to my knowledge" because as sure as I say all cellular phones definitely use pushbutton dials, I'll get a pointed letter from the Granny Bumpkin Patch Quilt and Cellular Phone Company informing me they carry a full line of pulse phones, and some even have hand cranks on the side. So, let's leave it at MOST cellular devices use DTMF.

Operation of the cellular system is merely an extension of the standard telephone structure. While cellular servers are, for the most part, independent companies, once the radio signals arrive, they are routed to the local central offices in the same fashion as conventional "land-line" calls.

The term "cellular" refers to a network of "cells" that cover a given area and pick up the radio signals. These cells overlap slightly, providing both comprehensive coverage of the area, and the option of handing off overflow calls in one cell to another cell. Today, better than 90 percent of the United States has cellular access. In all likelihood, those areas that don't have it probably don't need it. At least, not yet.

As for a cell, it is nothing more than a fancy repeater station. Phone calls from the mobile units are received in the 825 to 845 megahertz range, while the return signals are sent at 870 to 890 megahertz. Within each of these "bands" there are numerous individual frequencies, and this furnishes interference-free communications with large volume capability.

Hence, the various mobile and cell signals can be swapped back and forth among the allocated frequencies. Another advantage to this approach is that it's hard to monitor these calls. This is because they regularly change frequency. An important point to note however, is that while it is illegal to monitor cellular phone conversations, that hasn't stopped some from trying.

This brings us to the "long-distance" systems. As previously noted, microwave substations and satellites handle the majority of long-distance service. However, we are seeing an increase in the use of "short-haul" (short-distance transmission) microwave stations in larger cities, and this probably will continue.

But for the most part, microwaves are used to send telephone calls over long expanses ("long-haul" links). Due to the super high frequencies used (4 to 11 gigahertz), the nature of the transmission is "line of sight"; that is, each substation along the link must be able to literally see the substations on either side of it. In this fashion, the signal can be relayed along in a kind of electronic "bucket brigade."

So, whenever Loretta places a call to me, which is long-distance, microwave stations play an important role in my receiving that call. After her call leaves the interexchange office, it will travel through a myriad of microwave substations to arrive at its destination.

How many substations is dependent on several factors, but both distance and the terrain between our cities rank among the most significant. On average, microwave stations are 20 to 25 miles apart. Due to the earth's curvature, line of sight is hard to maintain much beyond 25 miles.

Of course, placing the microwave towers as high as possible (atop tall buildings and/or hills and mountains) does help increase the range, but such assistance is limited. That ol' earth is still curving and eventually overcomes even the tallest building or mountain.

The lay of the land, or terrain, can also play havoc with SHF microwave signals. They penetrate many of the materials we live and work around fairly well but are no match for dense forests, large metal/concrete structures and tall mountains. This type of terrain will both block and absorb the radio energy.

On the other hand, unobstructed flat land and bodies of water are easy marks for the system. Thus, when engineers are planning a microwave link, they naturally try to place the substations in areas separated by such terrain. Needless to say, when tall mountains are encountered, putting the station on top of them is a definite winner.

So, why is microwave so appealing for long-distance service? You might think that these complex substations, with massive and expensive towers, would be cost-prohibitive. But, when you consider the cost of copper coaxial cable, the labor to install it, the labor to maintain it, not to mention the expense of securing "right-of-ways" for the entire length of the cable, often the microwave approach is considerably cheaper.

In the final analysis, radio transmission has virtually replaced cable for continental long-distance service. And it's also employed for intercontinental service, although the mechanics are a bit different.

In this scenario, the now-familiar satellite is put into service. By positioning one of these celestial robots over the equator at an altitude of 22,300 miles (35,900 kilometers), it appears to be stationary. Actually, it isn't, but because of the altitude, it takes exactly 24 hours to orbit the earth; thus, the satellite is synchronized with the earth's own rotation and ends up in a "geostationary" orbit.

Geostationary orbits fall into the astronomical category of "geosynchronous" orbits which allow satellites to be positioned in a more or less permanent place — or, should I say, permanent *path*, as weather satellites use a form of this technique. In their case, instead of following the equator, they travel around the earth's poles. This results in a vertical survey of the planet several times a day.

Well, so much for the astrophysics lesson. What is important in all this is that communications satellites, utilizing the proper altitude, can be placed in a reasonably fixed position. Interestingly enough, this phenomena was first proposed back in 1945 by science fiction author Arthur C. Clark.

Anyway, with the satellites in place, ground stations can easily transmit data to and receive data back from them. Naturally, they are line-of-sight, so that means microwave frequencies (4 to 14 gigahertz) can be employed.

Atmospheric conditions, such as clouds and rain, have little effect on the radio signals; that is, until you get to about 14 gigahertz and above. In this frequency range, heavy rain can cause reflection of the signals — alas, poor reception.

Consequently, satellites are a natural for telephone communications. Ground associated problems, such as curvature and terrain, are irrelevant, and very few atmospheric conditions hinder their performance. But, they are expensive. No, let's make that EXPENSIVE!!!

Even though prices for both production and launching have decreased, mostly due to competition, you're still talking in the hundreds of millions of dollars to install a complete system. While there is little maintenance (I mean, it's hard to drive a repair truck up to the satellite. You would need a space shuttle.), they do have a life expectancy far shorter than the average microwave station, or central switching office.

However, once the initial investment is made, these flying phone booths tend to pay for themselves — at least, over the long haul. Hence, while still not used as extensively as other methods, satellites are creeping up. Aside from maintenance-ridden undersea cable or interference-ridden shortwave radio, they are the only viable means of getting telephone calls across the oceans. For that alone, they will endure.

As to how they work, think of them as "out of this world" (sorry) microwave stations. They sit up there and take the signals in, then relay them to wherever it is they are supposed to go. And, that should sound familiar.

More specifically, once your call reaches the central office, it is then transferred to an "earth station," either by cable or microwaves. The earth station acts as both an "uplink" and "downlink" by first transmitting the call up to the satellite, then receiving the response. There is, of course, another earth station at the other end.

The satellite itself has devices on board called "transponders." Their job is to manage all this back-and-forth communications. This is normally done by transmitting and receiving on different frequencies.

That technique preserves the "full duplex" feature (being able to talk and hear the other side at the same time) we have become so accustomed to with conventional telephone service. Some early "radio telephone" systems only allowed one side to talk at a time.

As can be seen, this is very similar to long-haul microwave. Some of the equipment is more sophisticated, and the concept is more glamorous, but the principle is about the same.

I have heard rumors, or should I say rumbles, about using lasers to transmit the signals instead of radio waves. But to date, they remain rumors. Personally, at least, I have not encountered additional details. However, considering the power of today's lasers, such a scheme should be more than feasible.

One last area to touch on in greater detail is fiber optics. A good deal more will be heard from this technology. "Lightwave" telephone systems have been around for more than a decade, and it isn't unrealistic to predict they may replace copper coaxial cable in local area service.

Fiber optics isn't as mysterious as sometimes thought. Basically, extremely high-grade glass is heated, then "pulled" into long and very fine fibers. These fibers have the ability to carry light over extended distances without appreciable degradation in signal quality.

With most optical fiber, this ability is the product of light bouncing off the interior walls of the glass. As it does so, it is "piped" down the length of the optical cable and emerges at the opposite end. Since the light can't escape (at least in theory), there is little, if any, loss. Hence, the signal arrives at its destination in virtually the same state as when it entered the cable. Slick, eh?

However, while theory is noble, and essential to research and development, it rarely is pure. Often, what looks sensational on paper ends up badly needing the services of the "ORKIN Man" when emerging as a prototype. And fiber optics is no exception. Some loss is experienced, usually due to light escape and glass absorption, so in a "long-haul" situation, amplification is necessary.

But, the signal quality received is often far superior to any other technique. For this reason, optical fiber is an important player in what has become a perpetual competition among phone services. Given a little more trial and error, it will establish its place in the realm of telephone communications.

## CONCLUSION

That pretty much covers operation of the telephone system, at least in the proverbial "nutshell." This, of course, doesn't delve into such things as bandwidths, markers, crossbars, originating registers, protocols, waveguides and the like. But unless you are planning a career in the telephone industry, you won't need to be concerned about that stuff.

To build the projects in this book, as well as design your own, this general overview should be more than sufficient. Actually, this degree of information is the perfect foundation, as one of the appealing aspects of any venture involving the telephone system is the revelations you encounter along the way.

And, on a final note; for us hobbyists, this is comparatively new territory. That, of course, always holds the allure of adventure, knowledge, and downright useful "gadgets."

# CHAPTER 2

# THE TELEPHONE LAB

If any project in this book deserves to be first, it is this one. I'm not positive, but maybe that's why I put it first. I don't know, uh..., it might have been for some other reason. Oh well, it's not important. Let's move on.

Seriously though, the Telephone Lab is a highly pragmatic workbench addition that will be of great assistance with both the remaining projects in this text and your future telephone "adventures." The lab provides a number of tools useful in designing and testing telephone-associated equipment, all packaged in a single handy unit.

Among those tools is a Dual-Tone Multifrequency (DTMF) generator ("encoder"), a DTMF receiver ("decoder") complete with LED readout and a digital voltmeter for measuring telephone line levels. Added to this is a multifunction line interface and a self-contained power supply.

Thus, this marvel furnishes an excellent workstation regarding telephone system endeavors. Paired up with the "Tool Box" in Chapter 8, you should have just about everything you need.

## THEORY

First take a look at *Figure 2-1*, the schematic diagram. Don't let this one scare you. Just because it spooks me doesn't mean it should throw you. It's really not as complicated as it looks.

Referring to the upper left corner, note the circuit starts with connection to the telephone line (tip/ring). To that we hook a panel neon lamp (it glows when a "ringer" signal is present), a binding post that provides front-panel line access, and a pair of RJ11 modular jacks, also mounted on the front panel.

The ring and tip then terminate at DPDT slide switch S1 that allows them to be sent to either a line interface or digital voltmeter (DVM). All of this furnishes several passageways to the incoming telephone line.

The interface circuit is a generic system that provides both rectified and nonrectified output. Metal-oxide varistor MOV1 protects against lightning-induced voltage spikes, while capacitor C1 blocks direct current (DC) that might be on the line. Transformer T1 matches the system to the standard 600-ohm impedance necessary for both line seizure and quality audio.

Diodes D1 and D2 are incorporated to "clamp" the alternating current (AC) "ringer" signal to about 3.9 volts. This, of course, helps prevent damage to voltage-sensitive components. Capacitor C2 is used to remove high-frequency harmonics that are either on the line itself, or caused by the diodes.

Single in-line pin socket SIP1 gives you access to an unrectified output, while bridge rectifier BR1 and SIP2 provide a polarity regulated source. Rectifying the incoming signal furnishes a fixed DC voltage (fixed in terms of poles) that is safe for integrated circuits (ICs) and/or other components where a change in polarity is problematic.

The other side of S1 feeds the input of a digital meter. This is a standard 2-volt configuration of the ICL7107 analog-to-digital converter (ADC). However, since the line voltage will range from 4 or 5 volts to as high as 200 volts (during telco testing), we need a meter that will read higher than two volts.

And that is where resistors R1 and R2 come into play. Arranged as a voltage divider, the resistors reduce potentials up to 200 volts to the 0-2 volts needed by U1. As for R3 and C3, these are normal input components for the 7107 circuitry.

The next section is the power supply. Made up of three regulators, it provides positive 5 and 12 volts and negative 5 volts. The ±5 is used for the analog converter, with +5 also powering the encoder (U5), decoder (U8), BCD decoder (U9), light-emitting diode (LED) display and buffer (U7).

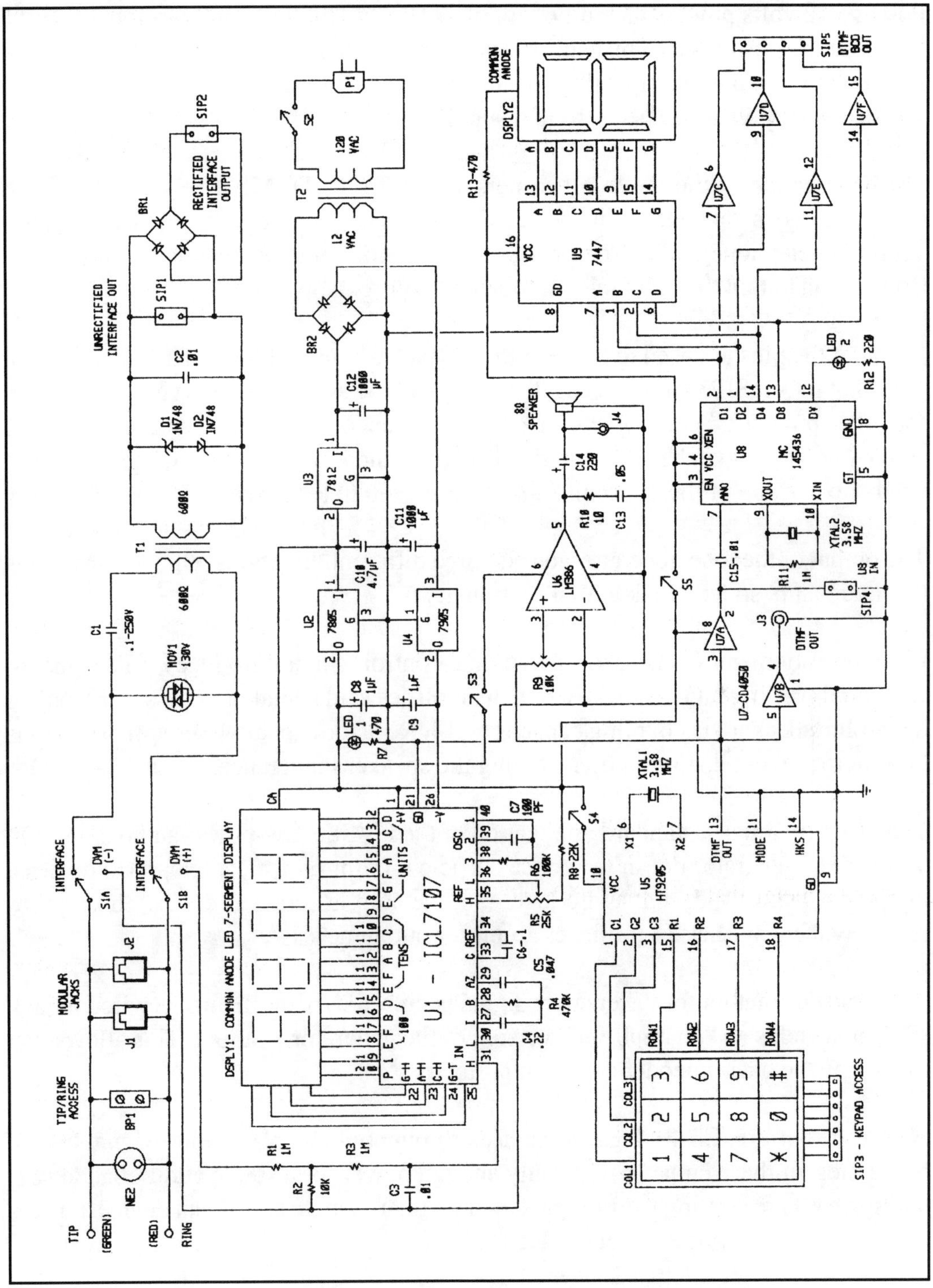

**Figure 2-1.** Telephone lab schematic.

Positive 12 volts powers the audio amplifier (U6). This chip will work off 5 volts, but its output power is diminished. Hence, the additional 12-volt potential is included. Incidentally, the buffer chip will also handle 12 volts, but since at these frequencies it works just as well at 5, I tied it to that supply.

The next section, in the lower left corner, is the HT9205 DTMF tone generator. Also referred to as an "encoder," this device (U5) turns strokes from the keypad into conventional telephone dialing tone pairs. Fairly simple in operation, it is crystal-controlled to maintain the strict telco frequency requirements.

Seven of the pins are used to connect the keypad, a standard X-Y matrix unit, to U5. Note that socket SIP3 allows access to the keypad rows and columns when it is needed for other projects. On an operational note, the encoder, decoder and amplifier all employ "on/off" switches (S3, S4 and S5). Since there will be times when you won't need one or more of these circuits, this permits you to turn them off.

The output of the tone generator goes in three different directions. One destination is the audio amp, so let's look at that section next.

Here, potentiometer R8 is used as a volume control which introduces U5's signal to U6's noninverting input (pin 3). U6 then amplifies said signal, and passes it along to the 8-ohm speaker via coupling capacitor C14. Resistor/capacitor network R9/C13 is present to prevent distortion by matching the speaker inductance.

That is pretty much a standard application of the LM386 low-power audio amplifier. The DTMF generator doesn't produce sufficient output for most telephone devices, or for that matter, a decent audio level. So, the amp is added to correct those problems. By the way, jack J4 furnishes amplifier output access.

A second destination for the encoder signal is buffer U7b. It, in turn, is wired to jack J3, which yields an "unamplified" version of the generator's output. This allows you to employ the basic encoder circuit in prototypes.

As for the buffers, CD4050 gates are used throughout the lab to boost signal levels. Since most of these chips provide only enough power for a single additional device, and since we are sending their outputs to more than a single additional device, buffers are necessary to service all needs. Hence, U7.

And, last but certainly not least, U5's output arrives at its third destination, buffer U7a. Here again, the signal is augmented before input to the MC145436 DTMF decoder U8. Note that U8's outputs are also buffered.

The decoder, or receiver, is another fairly simple circuit. It, too, is crystal-controlled, and produces a four-"bit" binary-coded decimal (BCD) output. Additionally, pin 12 provides a data valid (DV) signal (goes high) that confirms, via LED2 and R12, detection of acceptable DTMF tone pairs.

The input, pin 7, is coupled to the incoming audio through capacitor C15. This blocks any DC that might sneak onto the line. SIP4 allows external input to the decoder, while switch S5 handles power for both U8 and our next subject, U9.

The BCD coming out of U8 is compatible with any device that employs binary inputs. In our case, we want a visual display of the digit being received by U8. Thus, a good choice is the transistor-to-transistor logic (TTL) 74LS47 decoder/driver.

It will accept the BCD signal, decode it and send it to a 7-segment common-anode LED display unit. That, of course, will tell us whether the digit was a "2," a "5," a "9," or whatever. As previously stated, the BCD is also buffered and sent to SIP5. This socket provides auxiliary access to U8's output.

So, with all this "stuff" hooked up and turned on, you can press a button on the keypad and be able to both hear and see the digit/tone being produced. But that's not all. Each of the encoder, decoder and amplifier outputs, as well as the decoder input, is willingly available to help you with whatever top-secret project you have underway.

## CONSTRUCTION

Needless to say, this is a lot of circuitry in one box, so that box has to be an appropriate size. Actually, there is more of a panel need than housing need, but as seen by *Figure 2-2*, the printed-circuit board (PCB) is relatively large. So, I chose an 8-inch by 10-inch sloped panel case. Specific measurements are 8 inches (202 mm) deep, 10 inches (253 mm) wide and between 1 1/16 inches (26 mm) high at the front and 4 inches (102 mm) at the rear.

# TELEPHONE LAB PARTS LIST

SEMICONDUCTORS:

| | |
|---|---|
| U1 | ICL7107 3 1/2-Digit A/D Converter |
| U2 | 7805T Positive 5V Regulator |
| U3 | 7812T Positive 12V Regulator |
| U4 | 7905T Negative 5V Regulator |
| U5 | HT9205A DTMF Encoder |
| U6 | LM386N-1 Low-Voltage Audio Amplifier |
| U7 | CD4050 Hex Buffer |
| U8 | MC145436 DTMF Decoder (JDR) |
| U9 | 74LS47 BCD-to-7-Segment Decoder/Driver |
| BR1 | 400 PIV Rectifier Bridge |
| BR2 | 100 PIV Rectifier Bridge |
| DSPLY1 | 4-Digit LED Common Anode 7-Segment Display |
| DSPLY2 | Common Anode LED 7-Segment Display |
| LED1 | Green T-1 3/4 Jumbo Light-Emitting Diode |
| LED2 | Orange T-1 3/4 Jumbo Light-Emitting Diode |
| MOV1 | 130V Metal-Oxide Varistor |
| D1, 2 | 1N748 3.9V Zener Diodes |
| XTAL1, 2 | 3.579545 MHz Color Burst Crystals |

RESISTORS:

| | |
|---|---|
| R1, 3, 11 | 1,000,000 Ohm 1/4W Resistors |
| R2 | 10,000 Ohm 1/4W Resistor |
| R4 | 470,000 Ohm 1/4W Resistor |
| R5 | 25,000 Ohm PCB Potentiometer |
| R6 | 100,000 Ohm 1/4W Resistor |
| R7, 13 | 470 Ohm 1/4W Resistors |
| R8 | 22,000 Ohm 1/4W Resistor |
| R9 | 10,000 Ohm Panel-Mounted Potentiometer |
| R10 | 10 Ohm 1/4W Resistor |
| R12 | 220 Ohm 1/4W Resistor |

(Continued Next Page)

CAPACITORS:

| | |
|---|---|
| C1 | 0.1 µF 250V Mylar/Metal Film Capacitor |
| C2, 3, 15 | 0.01 µF Mylar Capacitors |
| C4 | 0.22 µF Mylar Capacitor |
| C5 | 0.047 µF Mylar Capacitor |
| C6 | 0.1 µF Mylar Capacitor |
| C7 | 100 pF Ceramic Disk Capacitor |
| C8, 9 | 1 µF Electrolytic Capacitor |
| C10 | 4.7 µF Electrolytic Capacitor |
| C11, 12 | 1,000 µF Electrolytic Capacitors |
| C13 | 0.05 µF Mylar Capacitor |
| C14 | 220 µF Electrolytic Capacitor |

OTHER COMPONENTS:

| | |
|---|---|
| NE2 | NE2 Panel Neon Lamp w/47,000 Ohm Dropping Resistor |
| BP1 | 2-Position Binding Post for Tip/Ring Connections |
| J1, 2 | 4-Position RJ11 Modular Telephone Jacks |
| J3, 4 | 1/8" Phone Plugs |
| T1 | 600-to-600 Ohm Matching Transformer (Jameco: 141621) |
| T2 | 120-to 12-VAC Power Transformer |
| SIP1, 2 | 2-Position SIP Socket |
| SIP3 | 7-Position SIP Socket |
| SIP4 | 4-Position SIP Socket |
| SPEAKER | 2" 8 Ohm Speaker |
| KEYPAD | 4-Row, 3-Column X/Y Matrix Telephone Keypad |
| S1 | DPDT Slide Switch |
| S2 | Push On/Push Off Power Switch |
| S3, 4, 5 | SPST Slide Switches |
| BRDBD | 2" by 3 5/8" Solderless Breadboard for Front Panel |

MISCELLANEOUS:

PCB Materials, Solder, Hardware, IC Sockets, Heat Sinks, Case (10" x 8" x 4" at back/1 1/8" at front sloping panel-RS), Hookup Wire, Paint, Dry Transfer Lettering, Epoxy, Etc.

This provides ample interior space, and while slightly cramped, enough panel space for the components used. The sloping panel is a real "plus" as it makes the unit far more convenient to work with, and that is always welcome.

OK, *Figure 2-2* is the PCB foil pattern, and *Figure 2-3* shows parts placement. I do recommend a PCB, but point-to-point wiring will work. Just be careful when soldering the ICs not to create any of those notorious solder bridges. That can really spoil your day, week, or year.

Start the assembly by installing all jumper wires and IC sockets. The small components, such as resistors, diodes and disk capacitors, come next, with the larger capacitors, rectifiers, crystals, transformer T1, MOV1 and regulators completing the "on-board" installation.

At this point, temporarily connect transformer T2, apply power and check PCB voltages as follows: U1 should show a positive 5 volts at pin 1, and negative 5 volts at pin 26. U3 will read positive 12 volts at pin 2, while U2 and U4 will show positive and negative 5 volt potentials, respectively, at pin 2.

U5 will have the +5 at pin 10, U7 at pin 14 and U8 at pin 16, while U6 will show +12 volts at pin 6. If all voltages check out, you at least know the power section is in order.

Set the PCB aside for the moment, and let's return to the enclosure. The case I used came with an aluminum panel, so I stashed that in a box, and cut a 1/8-inch piece of plexiglass to fit the panel opening. The plastic is far easier to work with, and you don't have to worry about accidentally shorting something out with the metal panel.

Appropriate holes and rectangular openings are cut, drilled, etc., in the panel, then it is sprayed with a coat of light-colored paint (I used light yellow). With the paint cured, dry transfer lettering is used to label. *Figure 2-4* shows the prototype front panel template which illustrates the designations I used (for whatever that's worth).

Follow this with a LIGHT coat of clear enamel to protect the fragile lettering. Let me stress LIGHT COAT! Dry transfer material, at least some of it, has a vicious habit of melting from the solvents in paint. Hence, a heavy coat of clear can easily result in a real mess.

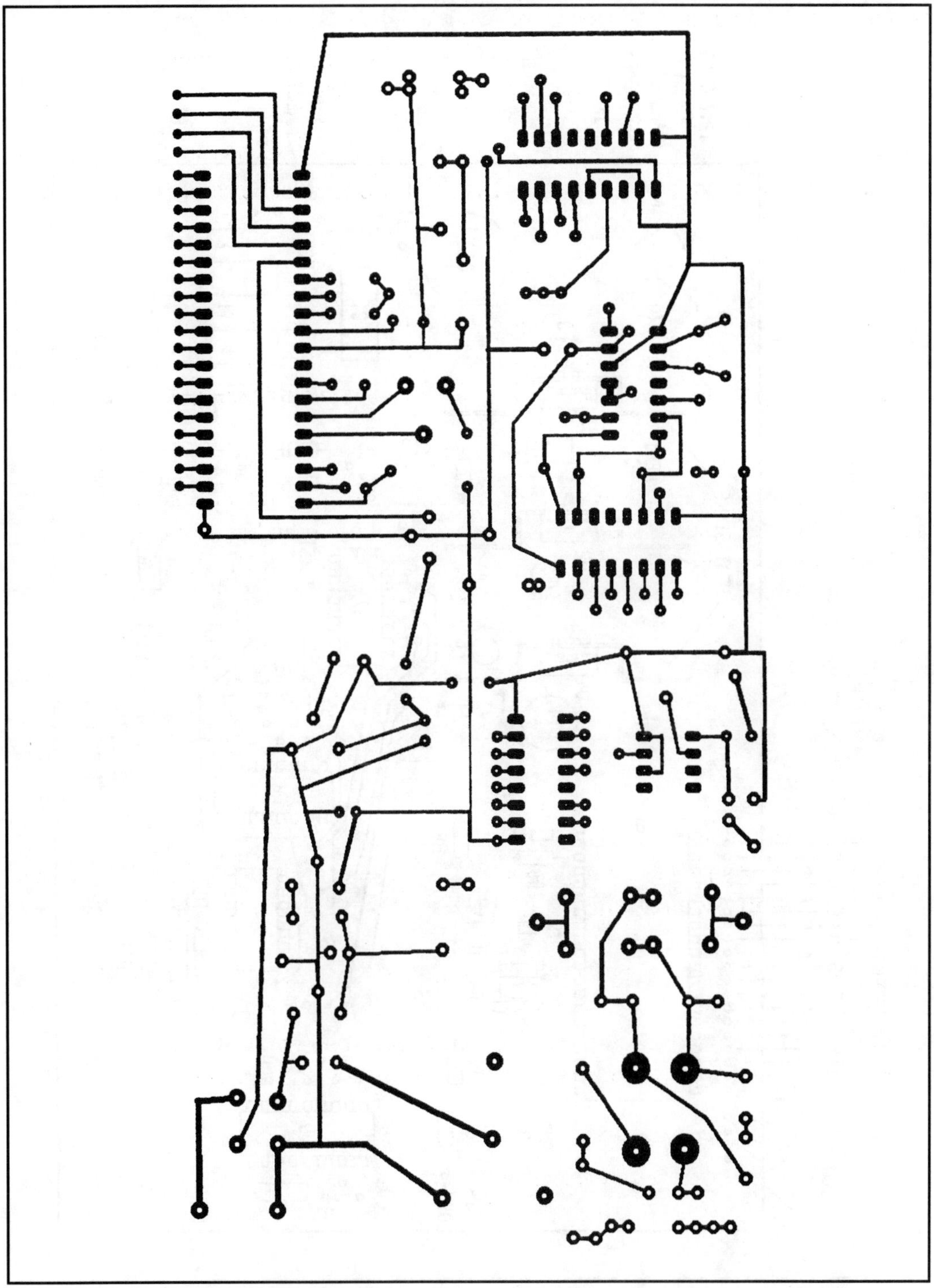

**Figure 2-2.** Telephone lab PCB foil pattern.

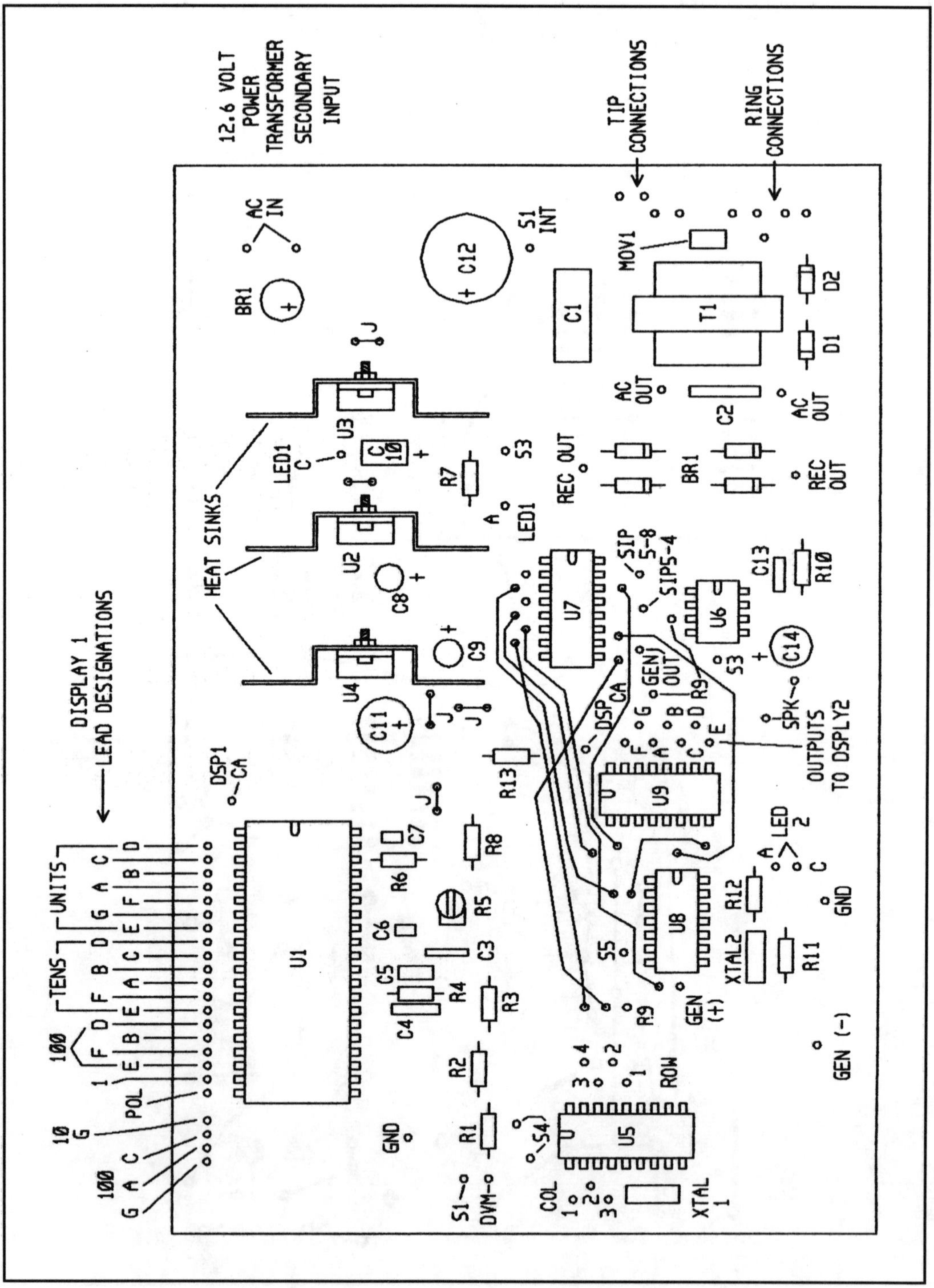


**Figure 2-3.** Telephone lab PCB parts placement.

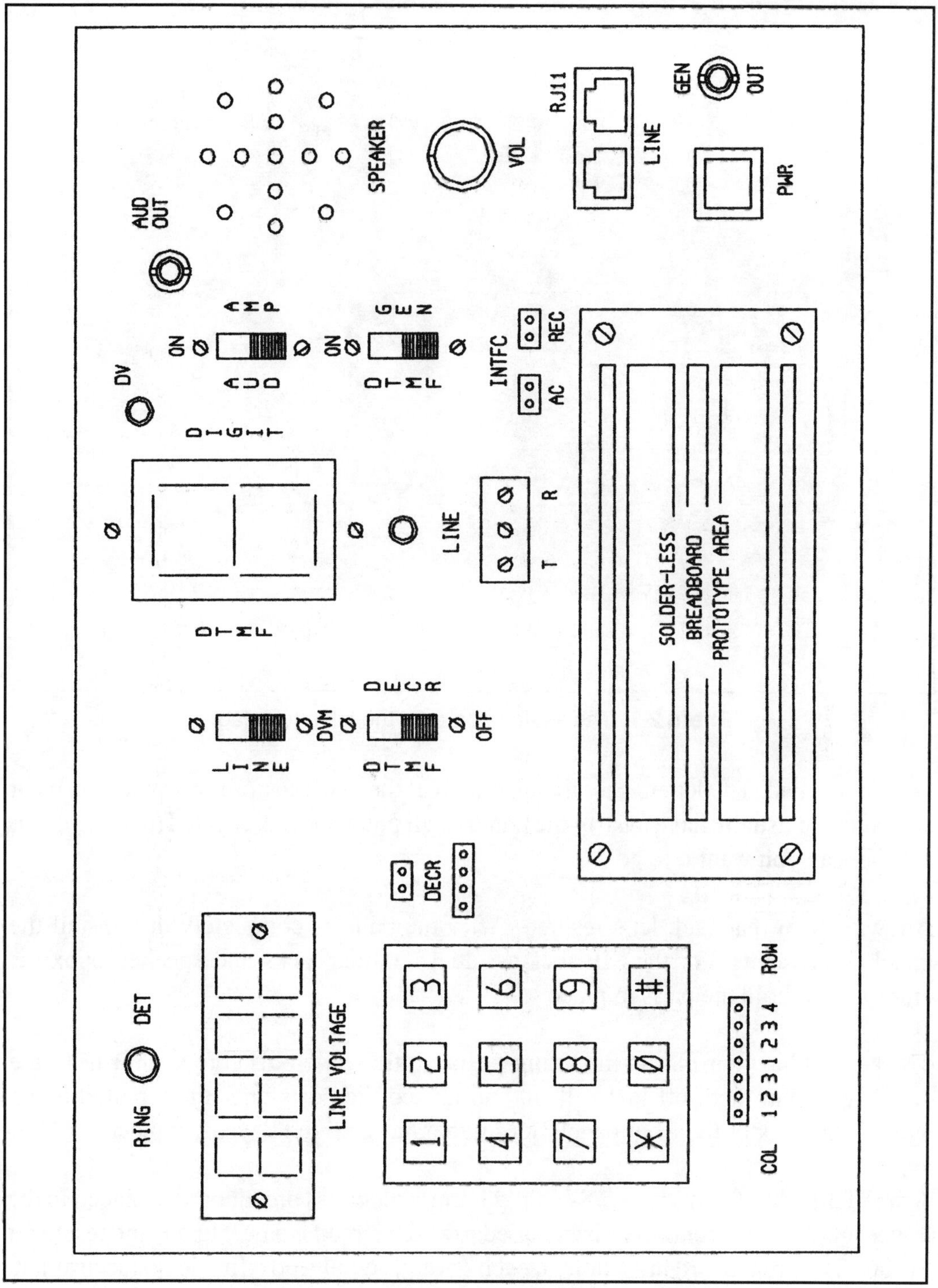


**Figure 2-4.** Telephone lab front panel template.

**Photo 2-1.** The stuffed board for the telephone lab.

I mean, it's bad. The letters become unreadable, the dry transfer runs under the clear coat and you usually have to strip the panel of all paint and start over. Trust me, this is not the way you want it to go!

Anyway, with that said, let's resume. After the paint is completely dry, install the panel components. For the SIP sockets, dual modular jacks and speaker, epoxy is employed to hold the parts in place.

The keypad has four plastic mounting posts, so these were used to anchor it. In the prototype, I cut them off to 1/4 inch, and heated the posts until they melted down against the back of the front panel. So far, that has held the keypad in place.

A small 2-inch (50 mm) by 3 5/8-inch (93 mm) solderless breadboard is placed in the lower center of the front. This is included to allow circuits to be laid out and tested on the lab. The breadboard has a hole in each corner, so bolts do a fine job of securing it.

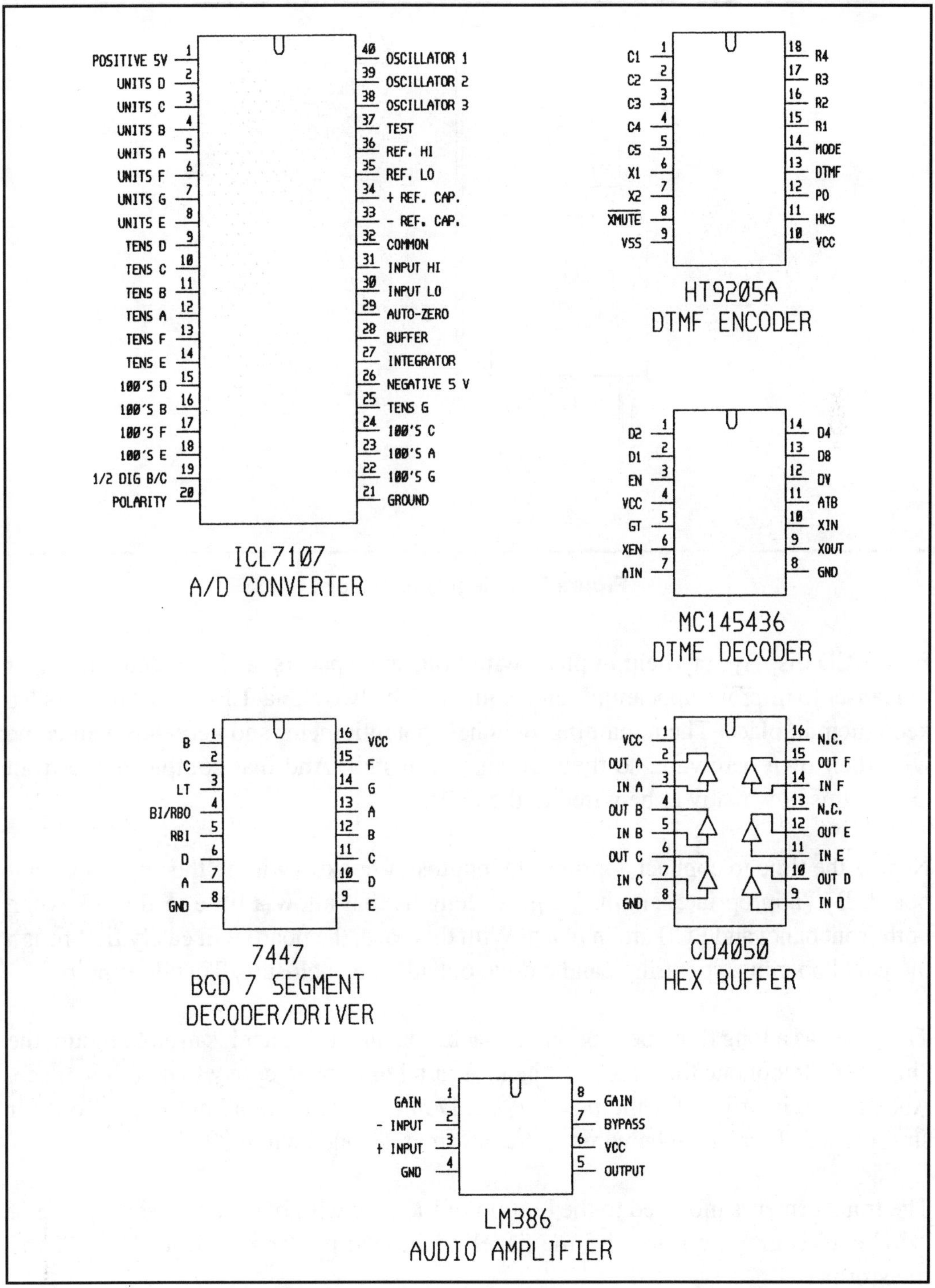


**Figure 2-5.** Telephone lab IC pinouts.

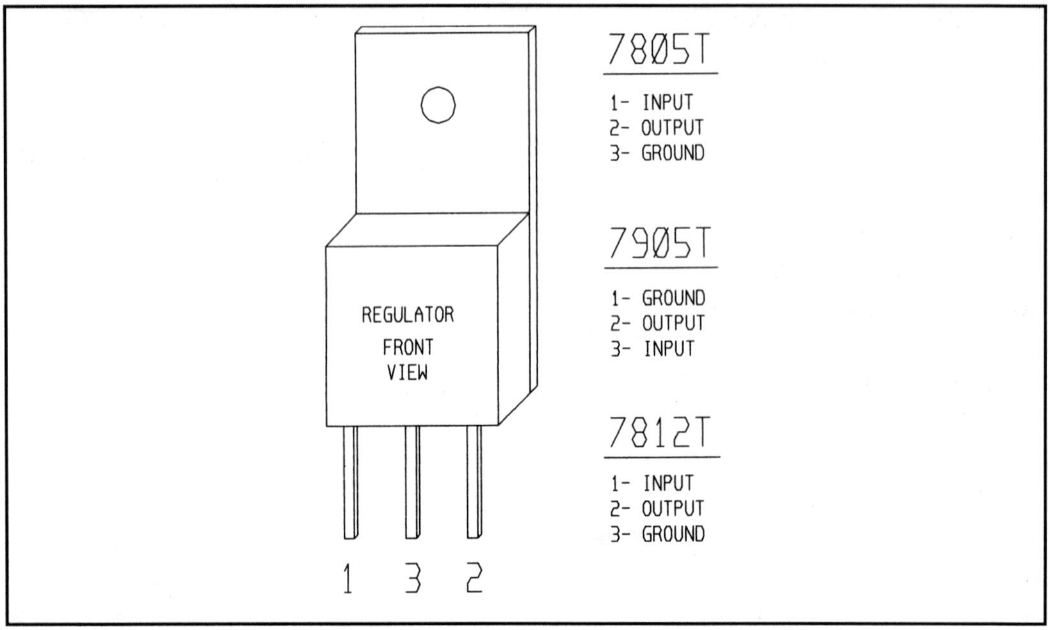


**Figure 2-6.** Regulator pinouts.

Both LED displays are held in place with bolts and spacers, and covered with clear red lenses to improve appearance and contrast. The two signal LEDs use "holders" to keep them in place. The remaining switches, potentiometer and neon lamp all come with their own hardware, so they are easy to install. And that completes the front panel. It is now ready to be wired to the PCB.

Now is the time to connect appropriate lengths of hookup wire to the printed-circuit board. By "appropriate lengths," I mean lengths that allow a little lead slack when both front panel and PCB are in place. With this case, the board will easily fit through the panel opening, so wiring can be done outside the enclosure. That does help.

The case has a long thin opening on the back side, and this panel is used to mount the "incoming telephone line" modular jack. Again I employed epoxy to hold it in place. Also, a hole is drilled for the power cord. Don't forget to tie a knot in the cord for strain relief. It can then be wired to transformer T2 and switch S2.

The transformer is mounted to the bottom of the case with bolts, and a short piece of "zip" cord is employed to wire the panel-positioned power switch to the cord and transformer.

Once all the wiring is completed, install the ICs in their sockets. As always, observe orientation, and be sure the pins seat correctly. A test is now in order.

## TESTING AND CALIBRATION

We have already run a test on the voltages, but that merely told us the power supply was working correctly and the IC sockets were receiving juice. Next, we want to confirm that all sections are cookin'.

First, turn "on" switches S2, S3, S4 and S5. That should light LED1, put a "0" on DSPLY2 and a "00.0" on DSPLY1. If not, double check the wiring. Also, if the DVM reads a little off "00.0," don't race around the room running your sentences together. The DVM hasn't yet been calibrated, which we will do in a moment.

Just kidding. I know you wouldn't do anything like that. All right, turn the volume control to mid-range, press any key on the pad, and you will hear the DTMF tones in the speaker. Furthermore, DSPLY2 will display the digit of the key you have activated.

If all that happens, then proper operation of the encoder, decoder and BCD decoder/driver is confirmed. A potential of close to 5 volts at SIP5 will also indicate the buffer is working.

Now, let's turn our attention to the other half of this device. Using a length of telephone cable with modular plugs on each end, install one in the RJ11 jack at the back of the lab, and the other in the wall modular jack.

First, flip S1 to the "DVM" position, and the digital meter will read somewhere in the neighborhood of 48 volts. What a neighborhood! I spent a month there one week. Never mind! Using a separate multimeter, check the voltage at BP1, the Tip/Ring binding post. Next, adjust PCB potentiometer R5 until the DVM matches the reading obtained with the multimeter.

That is the simplest way to calibrate the panel voltmeter. However, if you are a purist, you can measure the voltage between U1 pins 35 and 36, and adjust R5 until you get a 1-volt reading. This procedure sets the "reference voltage" to the recommended 1/2 full-scale potential.

As stated, this measurement should read about 48 volts. But, that is going to depend on your local telco and your individual telephone service needs. I've got so much stuff on my line that the telco jacked the battery voltage up to 50 volts. This was in direct response to my request to increase the line Ringer Equivalence Number (REN) level.

Also, some central offices use slightly lower voltages (46 or 44 volts for example), so you have to play that by ear. Take the independent reading with you DMM, and hope it is accurate.

To test the panel RJ11 jacks, plug a telephone, or other related device, into one of the jacks and try it. If you get a dial tone, that jack functions. Try the other jack, and if both work, you're in business.

Now, put S1 in the "INT" position. This directs the incoming line to the interface circuit. Test Both SIP1 and SIP2 with your multimeter, and you should see about 6 volts. Also, reversing the leads on SIP2 will change the polarity indication on the meter's display. That indicates the rectifier is truckin'.

As for the "ring detector" (NE2), have someone dial the line to which the lab is connected. Each time the phone rings, the neon lamp will glow, signifying the presence of the 90-volt AC "ringer" signal.

Another way to test this is to use the "Telephone Tool Box" in Chapter 8. That device includes a "ring generator" that produces the 90 volts necessary to trigger the ring circuit; hence, it will also light the neon lamp.

## OPERATION

How you use the telephone lab will depend largely on your projects. As seen, the unit encompasses a variety of "tools" that help make telephone prototyping much easier. So, you will have the final say on how those tools are put into service.

Let's say you are working on a circuit that requires BCD outputs of DTMF tone pairs. The encoder/decoder combination is your boy. If, however, your target is a device that demands telephone line interface, that part of the design is already here.

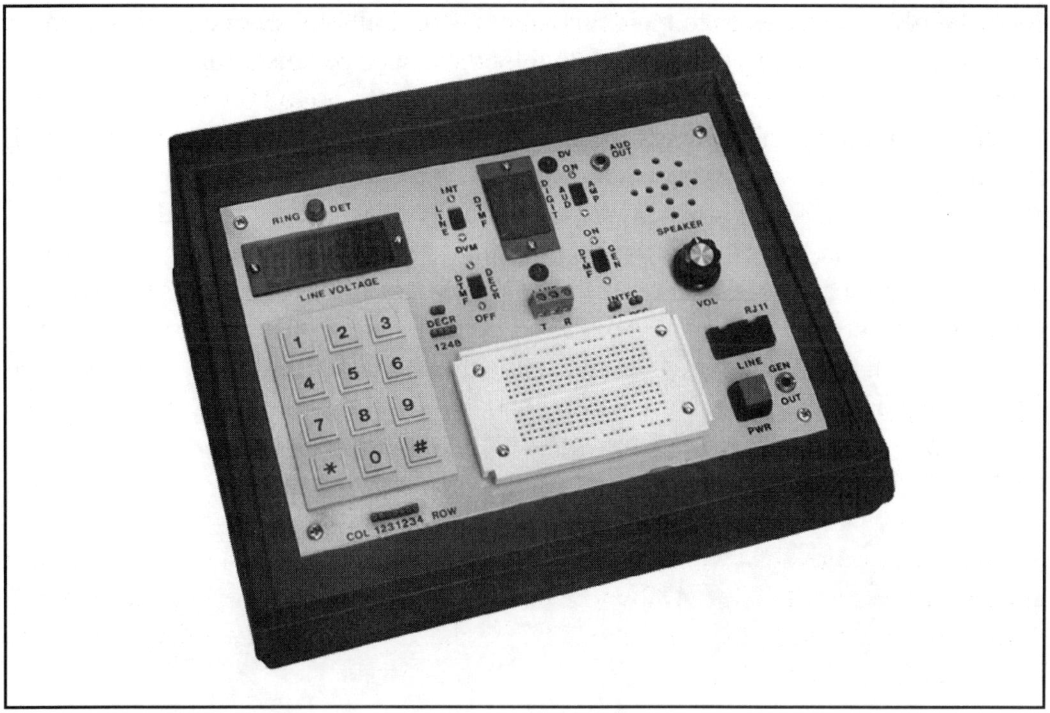


**Photo 2-2.** The completed telephone lab, ready for service.

And, of course, the solderless breadboard allows for testing, experimenting or otherwise "monkeying around" with new circuits or sections of a prototype. As I said, much of this is up to you.

As you work with the lab, its attributes will become all the more obvious. And that will lead to an ever-increasing ease of operation. In essence, don't worry about managing this device, as it practically does that itself. Your main concern will be coming up with telephone projects the lab can assist you with.

## CONCLUSION

All right, how's that for a first chapter? Build this gadget! You will like it! Even if you don't plan to work on any telephone projects, build it anyway. You'll like it that much!

However, once you have the lab on board, I suspect you will do some phone projects. Really, this device takes a lot of the "sting" out of telephone-related electronics. Now, that statement is not meant to bash this arena. Not at all. Telephone projects can be great fun and very rewarding. It's just that they can also be frustrating.

Much like video, this field tends to be temperamental. And, since you are dealing with the phone company, extra care has to be exercised so as not to cause grievous harm to the central office or its employees. They really get testy about that.

But, the telephone lab, and the rest of this text, comes to the rescue. Using the marvels, insider secrets and electronic wizardry contain herein, that mysterious world of telephones will be at your mercy. And, leading the charge against this foe will be...you guessed it...THE TELEPHONE LAB!

Enjoy!

# CHAPTER 3

# A 16-CHANNEL DTMF CONTROLLER

Here is a device that intercepts Dual-Tone Multifrequency (DTMF) tone pairs, and activates one of sixteen outputs. Each output can be connected to a relay, light-emitting diode (LED), silicon controlled rectifier (SCR), alarm or any other device you want to control.

And, the system isn't really particular where it gets those DTMF tones. They can come from the telephone line or any audio source, such as a receiver or tone generator — just as long as they are telephone company tolerance signals.

Hence, the possibilities for this unit are governed only by your ingenuity and imagination. Probably the most prominent application that comes to mind is remote control. With this device in place, virtually anything you would like turned on, managed, or otherwise manipulated can be with a simple telephone call.

That, of course, can encompass a wealth of different subjects. You could light up the house, turn on the air conditioner, turn off the oven, or adjust a security system, just to name a few, all by merely dialing home.

Naturally, the device has applications in a myriad of areas, and I will address some of those later. But for the moment, let's look at how the system works, and how to put it together.

## THEORY

As you may recall from the last chapter, the Motorola MC145436 DTMF receiver has the capability of decoding telephone dialing tones and converting them to binary coded decimal (BCD) outputs. For example, when you press the digit "5" on a "touch-tone" phone, you transmit the frequency pair of 770 and 1336 hertz.

The 145436 receives those tones and outputs the four-bit binary code of "0101"; that is, the D8 output is low, D4 is high, D2 is low and D1 is high. When you add D4 (4) and D1 (1), you should come up with 5. You do, don't you? OK, good!

I mean, I used to come up with 7, but then someone showed me the error of my ways. So, now I usually arrive at 5. At least most of the time. You know, I normally get, uh....5.

Yeah! Right! Enough of that! With that short refresher on the DTMF receiver under our belts, let's go back to the beginning and follow the circuit all the way through, referring to *Figure 3-1*.

Power is furnished by a standard 7805T (U1) linear regulated supply that I'm sure everyone is familiar with. A direct current (DC) voltage of more than 6, but not more than 35 volts, is applied to U1's input (pin 1). Capacitor C1 filters the input, while C2 serves to both filter the output (pin 2) and prevent U1 from oscillating. Linear regulators will sometimes do that.

The input to the DTMF receiver comes from either the telephone line, through the interface, or from an audio source. As for a straight audio input, this can be any device that produces speaker-level signals.

Regarding the telephone hookup, I employed an interface similar to the one on board the Telephone Lab. The line is sent to a matching transformer (T1) through capacitor C4. MOV1, a metal-oxide varistor, is used to suppress "on-line" voltage spikes.

The other side of T1 is wired in parallel with a zener diode network (D1 and D2) which "clamps" the alternating current (AC) ringer signal. Since the "ringer" is about 90 volts, this protects the circuit from that potentially damaging voltage.

Additionally, one leg of the secondary of T1 is connected to ground, and the other leg becomes the U2 audio signal. In either case, the audio is sent to U2 through coupling capacitor C3. This capacitor, by acting as a DC block, also prevents the normal 48-volt battery voltage from frying the receiver chip.

The crystal (X1) is a standard 3.579545-megahertz "color burst" quartz device, and it sets the timing for the MC145436. Note that a 10-megohm resistor is wired in parallel with X1 to stabilize the IC's oscillator circuit.

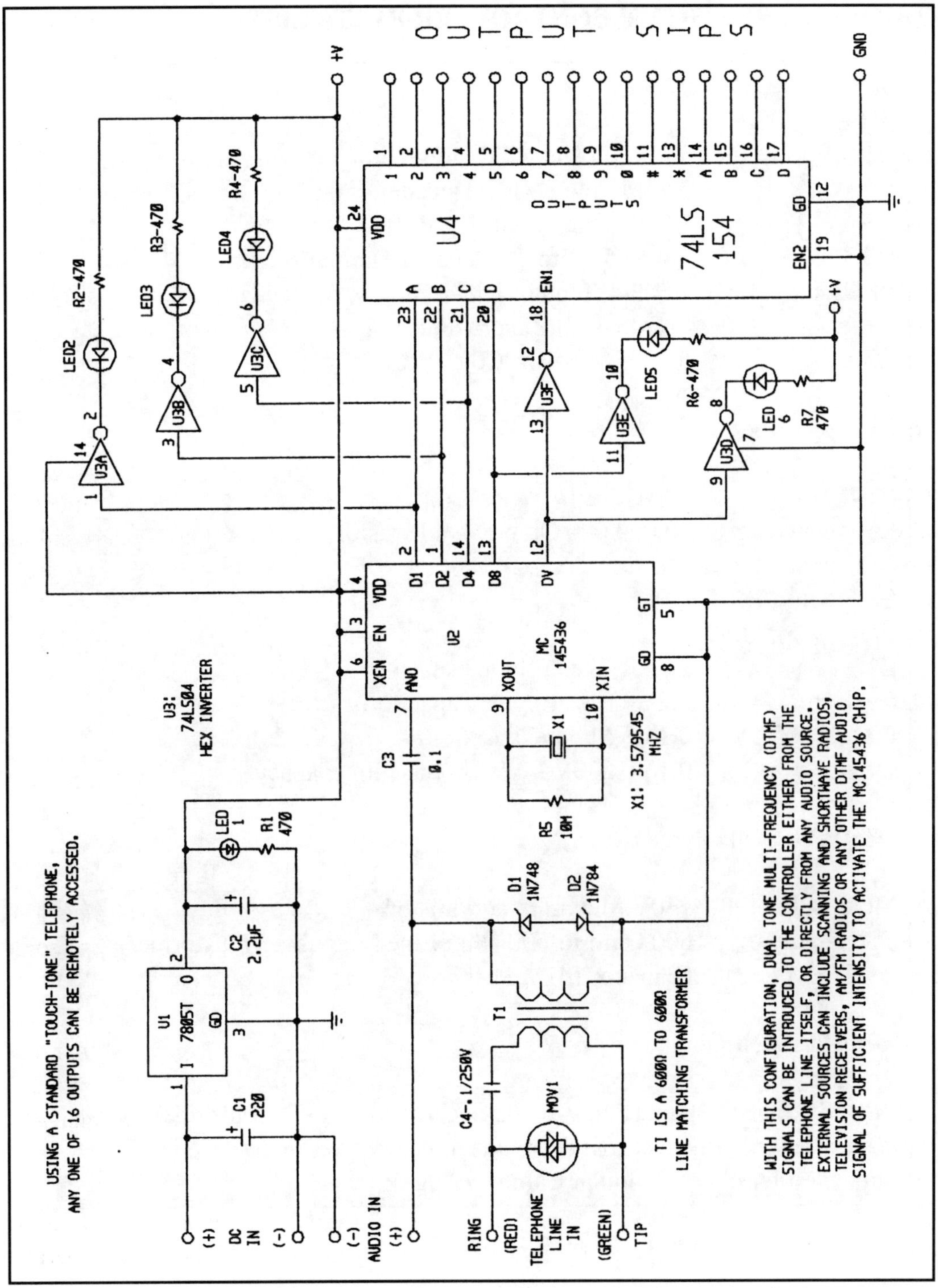


**Figure 3-1.** 16-output DTMF controller.

## DTMF CONTROLLER PARTS LIST

SEMICONDUCTORS:

| | |
|---|---|
| U1 | 7805T Positive 12V Regulator |
| U2 | MC145436 DTMF Decoder |
| U3 | 74LS04 Hex Inverting Buffer |
| U4 | 74LS154 4-to-16 Decoder/Demultiplexer |
| LED1-6 | T-1 3/4 Jumbo Light-Emitting Diodes (any color) |
| D1, 2 | 1N748 3.9V Zener Diodes |
| XTAL | 3.579545 MHz "Color Burst" Crystal |

RESISTORS:

| | |
|---|---|
| R1-4, 6, 7 | 470 Ohm 1/4W Resistors |
| R5 | 10,000,000 Ohm 1/4W Resistor |

CAPACITORS:

| | |
|---|---|
| C1 | 220 µF Electrolytic Capacitor |
| C2 | 2.2 µF Electrolytic Capacitor |
| C3 | 0.1 µF Mylar Capacitor |
| C4 | 0.1 µF 250V Mylar/Metal Film Capacitor |

OTHER COMPONENTS:

| | |
|---|---|
| MOV1 | 130V Metal-Oxide Varistor |
| T1 | 600 Ohm-to-600 Ohm Line Matching Transformer (Jameco: 141621) |

MISCELLANEOUS:

PCB Materials, Solder, Hardware, IC Sockets (14-pin, 24-pin wide), 16-pin header (output), Enclosure (if desired), Hookup Wire, Audio Jack, DC Power Supply (at least 6V), Modular Cable w/Plug, Etc.

All right! Now comes the fun. Once the receiver has converted the incoming DTMF tones to a BCD "nibble," what next? Well, that's a very good question. I'm glad I thought of it. Ha, Ha!

No! Uh...those four bits of data are now sent to the input of a 74LS154, 4-to-16 decoder/demultiplexer. This chip uses a BCD input to activate one of sixteen individual outputs. And, it is those outputs that provide the control capability.

When I say "activate," I mean that each of the sixteen outputs will go low when selected by the BCD demultiplexer. Thus, if you tie the other side of a load (relay, LED, etc.) to high, when the 74LS154 output goes low, the circuit is completed.

In this fashion, virtually any "load" can be controlled by DTMF tone pairs. Some, such as relays, may require driver circuits, but usually a single transistor or SCR does the trick. Remember, relay contacts can switch operating voltages on and off, and in turn, manage all sorts of external devices.

Also connected to the BCD outputs are inverting buffers (U3a-U3d). These are used to provide sufficient power to light indicators LED1 through LED4. This is a "passive" feature, and could be eliminated from the circuit if desired. But, they do tell you what binary code is being sent, and to that degree, are handy.

Note that a fifth inverting buffer/LED network (U3e/LED5) is connected to the "data valid" (DV) line of U2. Here again, this is a nonessential feature that indicates a valid tone pair, and could be left off. However, it is nice to know if the 145436 is receiving actual tone pairs and not just "noise."

There is one more inverter on the 74LS04 chip, and it handles the "enable" pin of the 74LS154. To activate the demultiplexer, the enable line has to be low, and since we are using U2's data valid (DV) signal to do the honors, we have a problem; that is, the DV goes high when a valid tone pair is received. However, that last inverter changes the high to a low, and the 74LS154 activates.

With that said, we are done with theory. Basically, this is a simple digital circuit that uses computer-related technology to perform a useful task. Gosh! What more could you ask for? Please, don't answer that!

## CONSTRUCTION

I used printed-circuit board (PCB) methods to construct this device, but then, I almost always use printed-circuit methods. It can be done just as easily with perf-board and point-to-point wiring. However, I recommend a PCB, so I've included *Figure 3-2*, a solder-side foil pattern, and *Figure 3-3*, a parts placement diagram, for your convenience.

Using the PCB, install the jumpers and integrated circuit (IC) sockets first. Then come the resistors, capacitors, crystal, MOV, LEDs and transformer. For the outputs, I used a 16-position single in-line pin (SIP) socket soldered to the PCB pads. Also, SIP sockets are used for the positive and ground connection points. Finally, install regulator U1. With this done, it's time to run a quick voltage test.

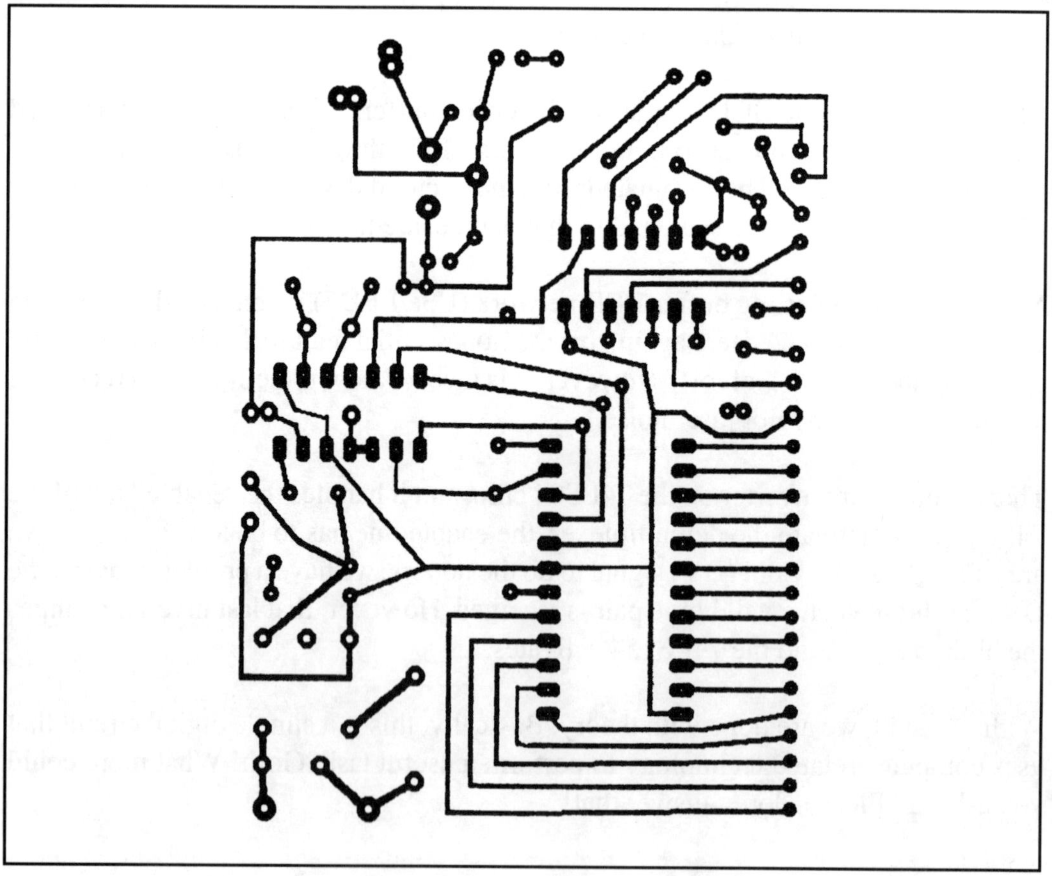

**Figure 3-2.** DTMF controller foil pattern.

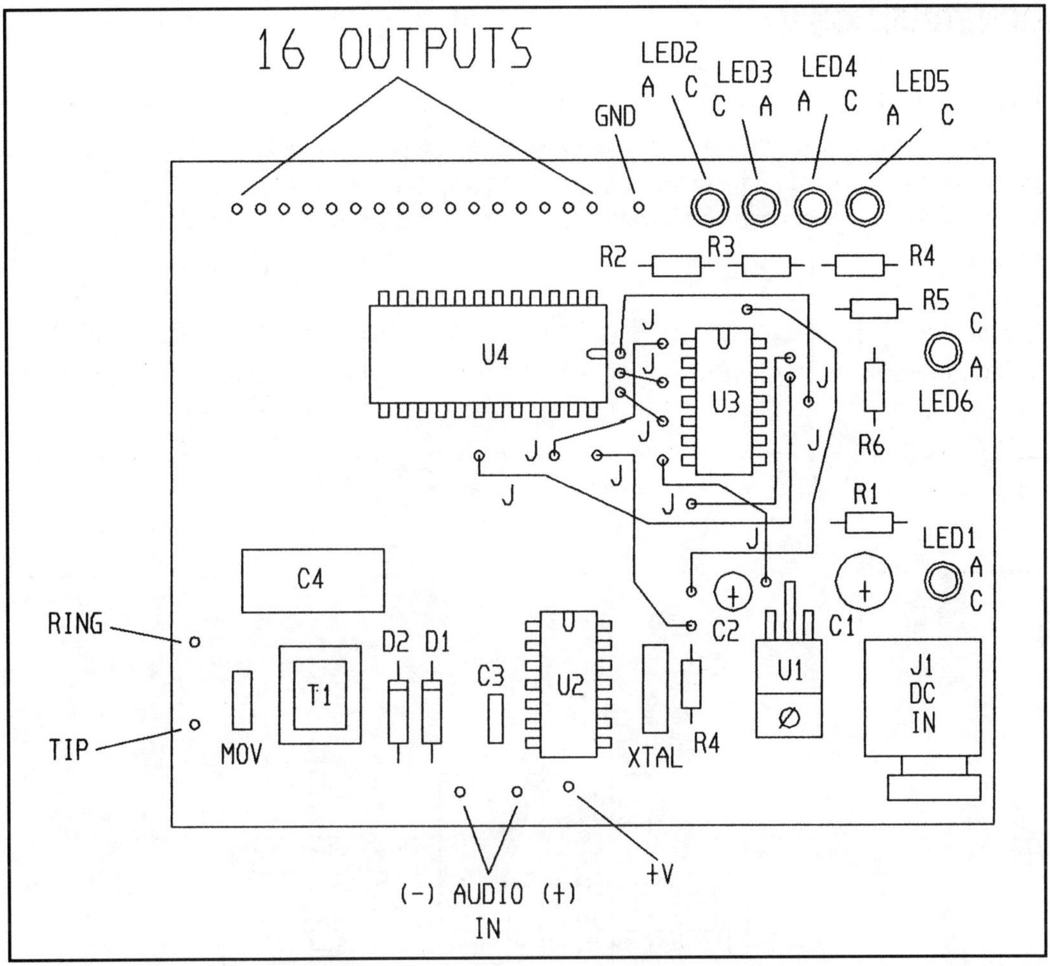


**Figure 3-3.** DTMF controller parts placement.

Connect your DC power source, and check the voltage at pin 2 of U1. It should be real close to positive 5 volts. If not, recheck the orientation of the IC and capacitors C1 and C2.

Next, test the voltage at pin 4 of U2, pin 14 of U3 and pin 24 of U4. Again, a +5 volts reading is expected. If all the voltage points check out, remove the DC source and install the chips.

The last assembly step depends on how you plan to use this device. For the prototype, I soldered short lengths of hookup wire to the "audio in" and "DC in" points, and a section of modular telephone cable to the "telephone line in" PCB pads.

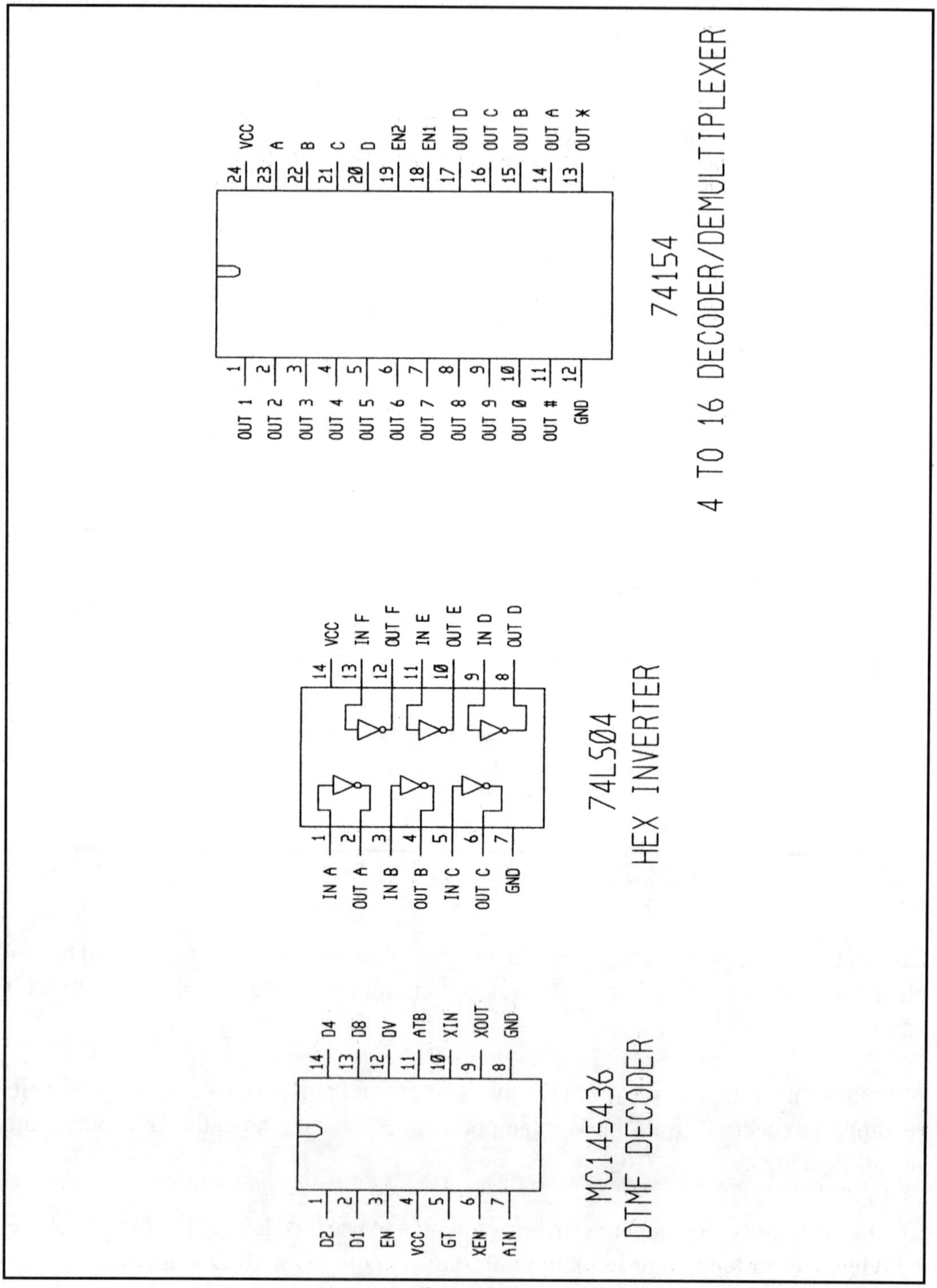


**Figure 3-4.** DTMF controller IC pinouts.

**Photo 3-1.** The stuffed board on the 16-input DTMF controller.

This provides a quasi-generic configuration that easily adapts to many different applications. However, if you have a specific mission in mind, now is the time to connect appropriate leads to the three inputs.

Once that is completed, the unit is ready to be tested, then put into service. And, at the risk of stating the obvious, that will be our next challenge.

## TESTING

To test this extraordinary specimen of electronic ingenuity, we will need a DTMF tone source. That's right! You guessed it. A "touch-tone" telephone is a superb choice. It's also handy.

With power connected to the circuit, insert the RJ11 plug into a modular wall jack. Next, hook a multimeter to one of the output SIP sockets and ground. You should see a reading of close to positive 5 volts.

Now, take a "tone" style telephone "off-hook" and press the button that corresponds to the output the meter is connected to. For example, if the multimeter is wired to U4 pin 6, you will want to press the "6" digit key. This should result in the meter dropping to a measurement of zero (low).

Try replacing the meter with an LED/470-ohm resistor network. Hook the LED's cathode to the output socket, wire the anode to the resistor and the other end of the resistor to the positive voltage socket. Now, pressing the proper key will light the diode.

It is best to check each of the outputs individually, just to be sure the 74LS154 is completely functional. While rare, I have encountered demultiplexers that had one or more faulty outputs. Such a circumstance can lead to a puzzling situation when the controller, for no apparent reason, suddenly fails to do its job.

If all is well, testing is completed. Let's now chat about putting this device to work. I think its versatility will amaze you. Well, maybe not amaze, however....maybe it will amaze you.

## OPERATION

As earlier stated, aside from just telephone tones, the unit can detect DTMF signals from just about any source that produces DTMF signals. Input directly to the MC145436 does have to be "line level," or strong enough to power a speaker, but such signals cover a lot of ground.

With this in mind, the controller becomes a valuable friend to the amateur radio operator (HAM). Since many of the present-day transmitters have DTMF encoders, HAMs can use the device to manage repeater stations or other gear.

This, of course, functions in the same fashion as taking the tones off the phone line. As long as the tone pair is a valid telephone system signal, the receiver will acknowledge it, and send the BCD equivalent to the decoder/demultiplexer.

Also, as earlier stated, DTMF tones received by scanners, televisions or shortwave radios can be decoded. Thus, with our device connected to the speaker terminals, or output jack of a scanner, TV, etc., as the signals come in, they will activate the various controller outputs.

While this won't exactly provide "control" in the classic sense, it can be used for "attention-getting" contraptions (advertising signs for example) and amusement. As the incoming signals rapidly change, so will the 74LS154 outputs. With these connected to LEDs or lamps, a veritable light show can be produced.

Getting back to more pragmatic applications, the device can be used to control up to sixteen home appliances or functions. Of course, with traditional telephones, only twelve outputs can be activated. You will need a phone with a fourth row (A, B, C, D) to access all sixteen lines.

Even so, twelve "whatevers" is a lot of control. For example, you could use "1" to turn on the porch light, "2" for the air conditioning and "3" for the living room. "4" could handle power to the bedrooms, while "5" controls the electric range. "6" and "7" might manage the security lights and alarm system, and "8" the garage lights.

So, as can be seen, twelve outputs will handle a multitude of needs, with several outputs to spare. And, all of this controlling can be done from any "touch-tone" telephone, anywhere in the world.

Let's now take a look at some of the "loads" that will be useful. *Figure 3-5* illustrates how to connect three different devices to the controller's outputs, with each example servicing a unique need.

The first scenario, *3-5a*, takes advantage of the transistor-to-transistor logic (TTL) output of the 74LS154. TTL chips provide enough power to light an LED, or for that matter, activate at least one additional logic device.

Since U4's outputs are high until triggered, the diode is connected, through a dropping resistor, to the positive voltage point (Vcc). Using this configuration, an activated output lights the LED because that output goes low and completes the circuit.

Example *3-5b* is a simple single-transistor driver circuit. In this instance, the load is a relay, but other loads can be substituted. Just don't exceed the transistor's power rating. If a higher current/voltage requirement is in order, go with a power transistor.

With this circuit, the potential applied to Q1's base, from the 74LS154, allows the positive voltage to pass from Vcc to ground. That, in turn, activates the relay. However, when the 154 output goes low, voltage flow stops, and the relay deactivates.

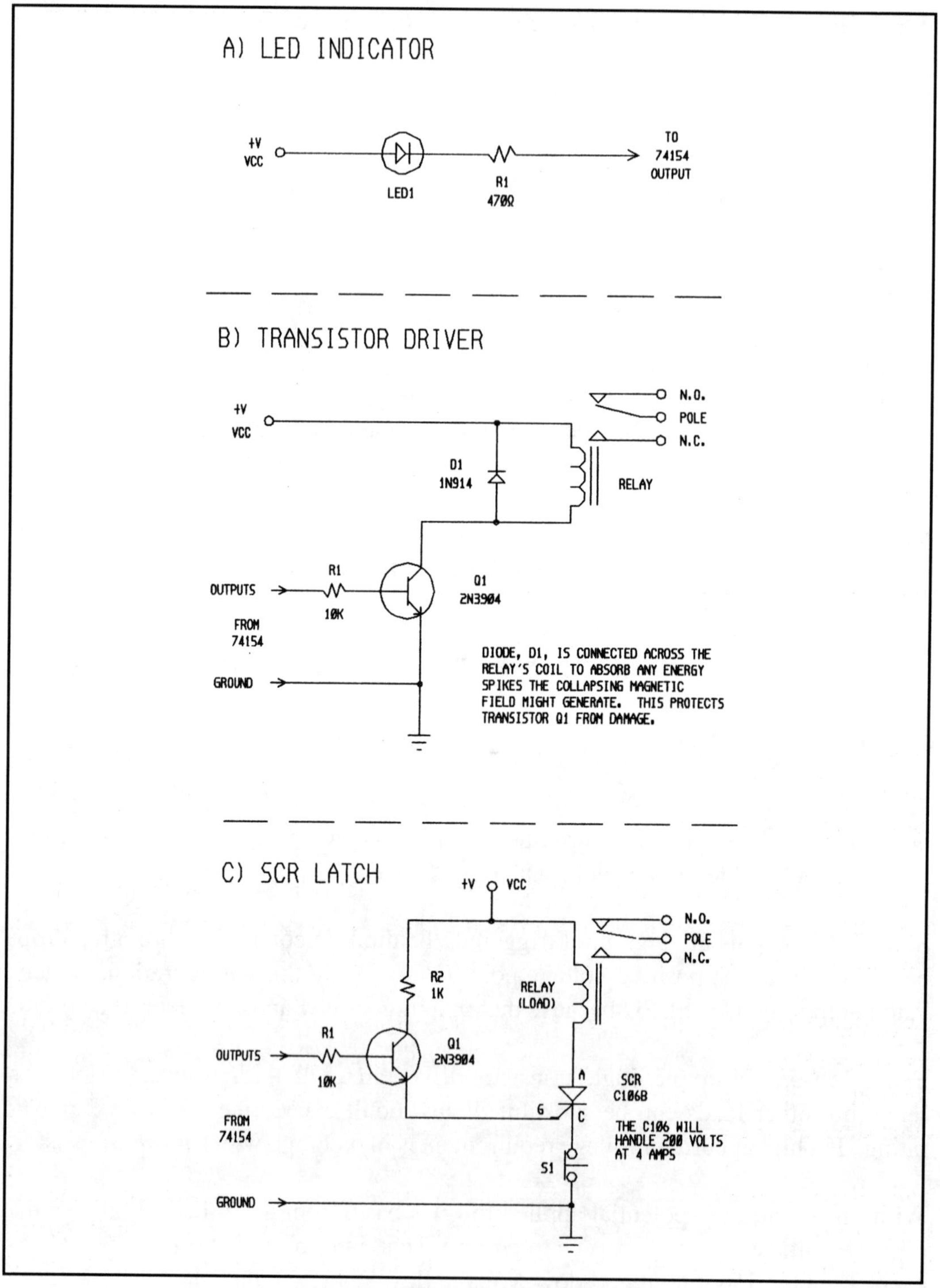


**Figure 3-5.** 74154 output control circuits.

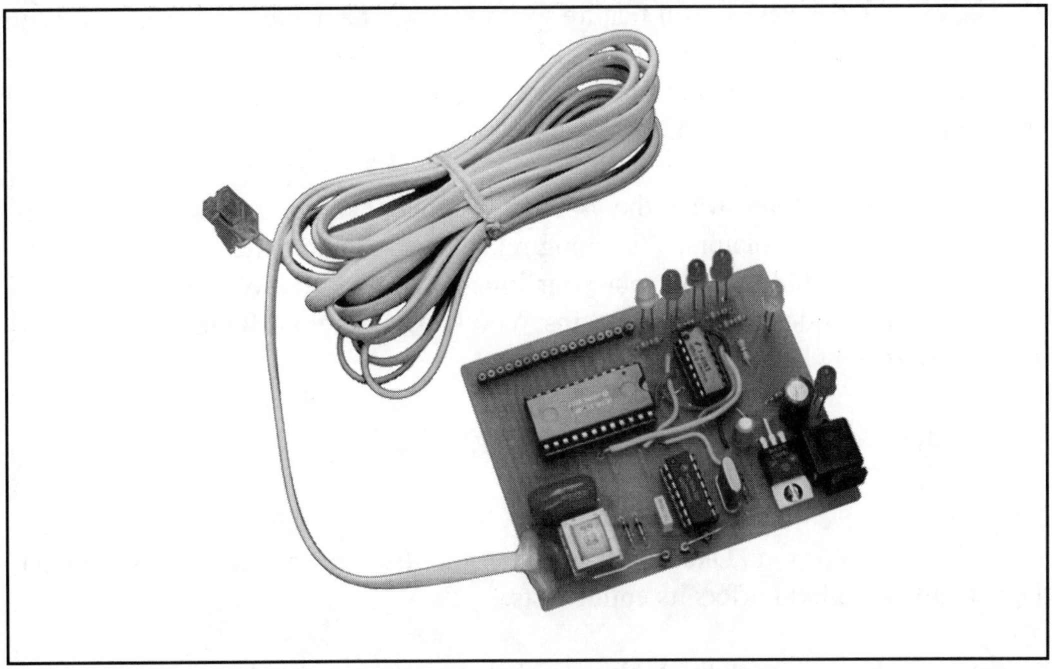

**Photo 3-2.** The completed 16-output DTMF controller. This circuit can be enclosed if desired.

Thus, a relay with SPDT contacts will provide both "normally open" (NO) and "normally closed" (NC) switching. This allows you to either turn something off, or turn something on, depending on which set of contacts you employ.

The last illustration, *3-5c*, uses a silicon controlled rectifier (SCR). With this system, a signal from U4 will latch the SCR, keeping whatever load connected to it activated until the circuit is reset (in this case, by pressing switch S1).

Since the SCR needs a high signal, and the 74LS154 produces a low, when triggered, transistor Q1 and friends (R1, R2) are required to invert the "high" output of the idle demultiplexer. Again, a relay is the load, but then relays are great for controlling other devices. Like *3-4b*, other loads can be used.

The silicon controlled rectifier is similar to the transistor, in that power applied to its gate (equivalent to the base) permits flow from the anode to the cathode (equivalent to the collector and emitter). The difference is the SCR latches into this state until another signal releases it.

This is, of course, a very useful feature when you want a device to remain on. With the transistor circuit, as soon as the DTMF signal is removed, the load returns to the initial state. Thus, it is good for momentary activation, but the SCR fills the bill when you want long-term activation.

Anyway, those are three ways the 74LS154 outputs can control external devices. Such things as "optoisolators," "latching relays" and "logic gates" can also be employed to do your bidding. So, put your imagination and the ol' "gray matter" to work, and the world is at your fingertips. You know, "let your fingers do the walking," or something like that.

## CONCLUSION

I hope you have enjoyed this chapter. I've had a lot of fun with this gadget, and probably will have more. One aspect of electronics that never ceases to amaze me is the versatility the field offers its enthusiasts.

So much can be done with the tools provided, and often, individual components far exceed their original intended application. Take the 74LS154, for example. This chip was initially developed as a way to direct data flow in computer systems.

However, recognizing its characteristics and capabilities, the 74LS154 can be put to use in a variety of non-computer system projects. And for me, finding a new application for a supposedly "dedicated task" IC is part of what makes electronics fun.

I trust it is for you, also. Every time I work with a new, or even an older IC, I study its functions, and try to visualize unique employment of the chip. It's amazing how often I do find new purposes for the little darlings. And, you can too.

# CHAPTER 4
# THE ELECTRONIC BELL

Here is a fairly simple project that will be useful in many of your endeavors. If you are building equipment that requires recognition of a telco "ringer" signal, this baby will do the job. Or, if you would like to have an extension "bell," say, out on the patio, again this project is for you.

Also, if you still own a phone with the old "bell and clanger" style annunciator, this circuit provides some modernization. It is small enough to easily fit inside such a phone, and brings the "beast" into the world of contemporary electronic ringing.

"Beast" — that's a terrible thing to call one of those dinosaurs. Well, I guess "dinosaur" isn't much better. They served their purpose for many years, but time marches on. Today, the "warbling" timbre of modern phones is by far more pleasant to the ear.

And, with three integrated circuits (ICs) and a handful of support components, you can add that sound to any telephone or project. With that bit of salesmanship out of the way, let's take a look at how this system does its job.

## THEORY

Serving as the heart of this device is the LM555 timer/oscillator. This chip, I'm certain, is an old friend to all, and is employed here in its "astable" configuration. Actually, there are two astable oscillators present, with one controlling an initial tone, and the other handling a second tone.

Hence, the "ringing" is comprised of two alternating tones that can produce a number of different sounds. I will get into this in greater detail later, but this aspect helps make the project fun.

Looking at *Figure 4-1*, it can be seen that the electronic bell consists of two sections: first, the telephone line interface; and second, the dual oscillators. The interface is used to detect a ringer signal, and the oscillators are, of course, used to produce the commotion.

Regarding the interface, the "tip" and "ring" are directed to the input, or light-emitting diode (LED) side, of an H11AA1 optoisolator (U1). Metal-oxide varistor MOV1 protects the device against on-line voltage spikes (often lightning induced), while resistor R1 drops the ringer signal to a safe level for U1.

When the "ringer" is present, U1's internal LED glows, triggering the phototransistor and allowing current to pass from collector to emitter. Since the H11AA1 is activated only by alternating current (AC), the normal direct current (DC) battery voltage has no effect on the optoisolator.

The current flow from the phototransistor side of U1 is applied to the base of transistor Q1 through dropping resistor R8. That base potential permits Q1 to transfer current, and in turn activate relay RLY1. Since the relay contacts are connected to the 12V power supply, 12V is now sent to the oscillator section, making it operational.

Network LED1/R7 is included to confirm the presence of the external 12-volt source, while LED2/R9 lets you know power has been delivered to the oscillators. Both are passive actions, and can be omitted if desired, but personally, I like to know that the "juice" is where it's supposed to be.

As for the oscillators, both are configured as astable timers. U2 uses resistors R2 and R3 and capacitor C1 or C2 as an R/C style tank circuit that sets the initial tone's duration and frequency. Note that switch S1 selects either C1 or C2. This permits you to change the sound of the "bell."

U3 operates in the same fashion, with R5, R6 and C3 determining the second tone's pitch and duration. Additionally, U2's output (pin 3) triggers U3 through resistor R4. That "triggering" occurs on a low signal that is part of U2's output cycle.

Hence, the tones are produced as an alternating, or "two-state" sequence that generates the "ringer" signal with the speaker. As to what that signal will sound like depends on the values of C1, C2, C3 and R4. I employed two capacitors in the first oscillator (U2). The 1-microfarad unit C1 produces that familiar "warble" pitch.

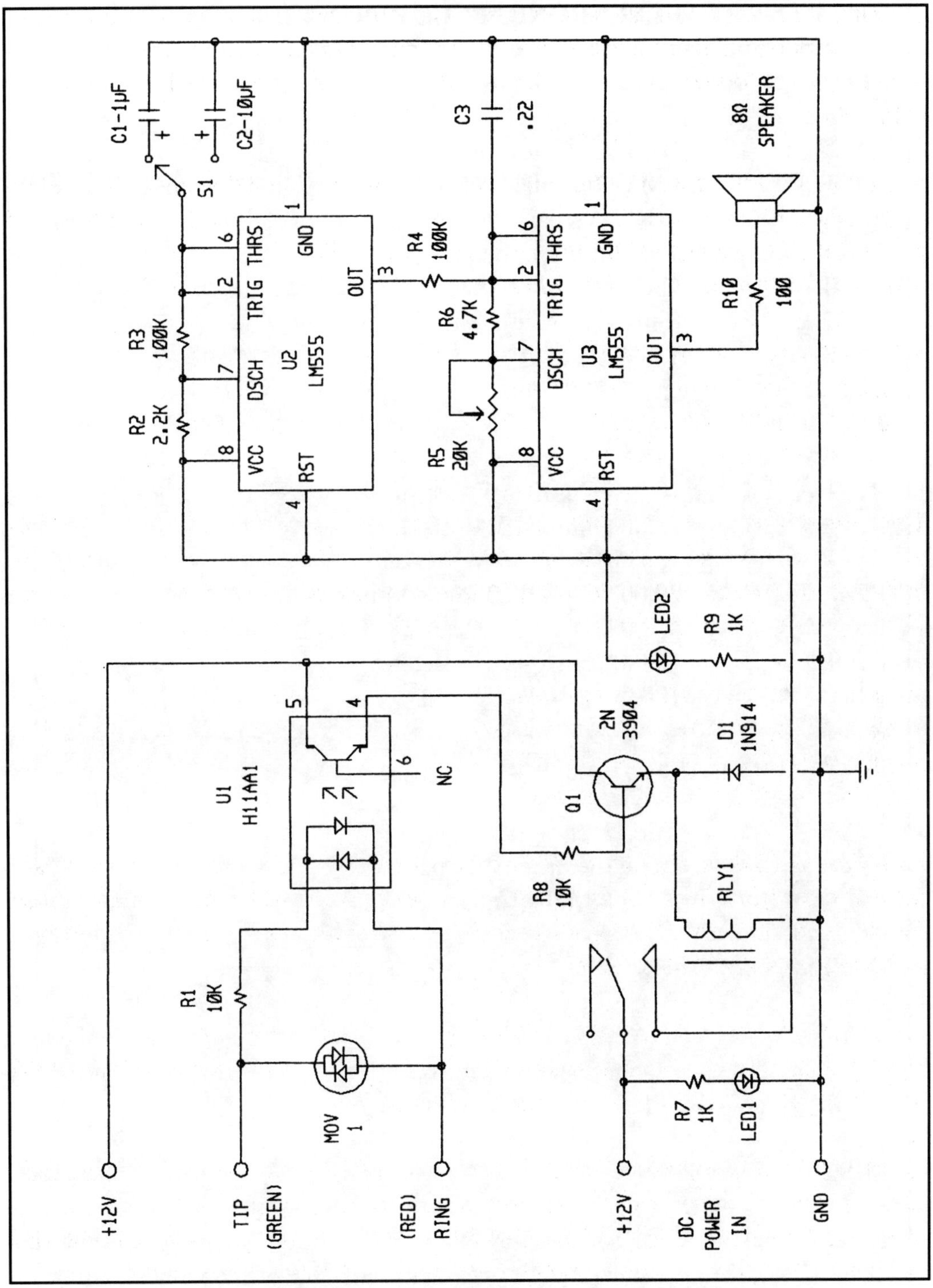


**Figure 4-1.** Electronic bell ringer circuit.

## ELECTRONIC BELL PARTS LIST

SEMICONDUCTORS:

| | |
|---|---|
| U1 | H11AA1 Optoisolator |
| U2, 3 | LM555 Oscillator/Timers |
| Q1 | 2N3904 NPN Signal Transistor |
| D1 | 1N914 Signal Diode |
| LED1 | Green T-1 3/4 Jumbo Light-Emitting Diode |
| LED2 | Red T-1 3/4 Jumbo Light-Emitting Diode |
| MOV1 | 130V Metal-Oxide Varistor |

RESISTORS:

| | |
|---|---|
| R1, 8 | 10,000 Ohm 1/4W Resistors |
| R2 | 2,200 Ohm 1/4W Resistor |
| R3, 4 | 100,000 Ohm 1/4W Resistors |
| R5 | 20,000 Ohm PCB Potentiometer |
| R6 | 4,700 Ohm 1/4W Resistor |
| R7, 9 | 1,000 Ohm 1/4W Resistors |
| R10 | 100 Ohm 1/4W Resistor |

CAPACITORS:

| | |
|---|---|
| C1 | 1 μF Electrolytic Capacitor |
| C2 | 10 μF Electrolytic Capacitor |
| C3 | 0.22 μF Mylar Capacitor |

OTHER COMPONENTS:

| | |
|---|---|
| RLY1 | 12V PCB Relay |
| SPKR | 8 Ohm Speaker (any size up to 4") |
| S1 | SPDT Slide, Toggle Switch |

MISCELLANEOUS:

PCB Materials, Solder, Hookup Wire, 5-to-15 VDC Power Supply, IC Sockets, Telephone Cable w/Modular Plug, Power Jack to Match Power Supply, Etc.

The 10-microfarad capacitor (C2) results in a sound much like a European police siren. Quite the attention getter. Either might be your choice, but the best way to ascertain which you prefer is to breadboard the circuit. That way, you can try various values for capacitors C1 and C2, as well as C3, and determine the "noise" you like best.

While you're experimenting with the capacitors, also try adjusting potentiometer R5. Its setting, too, has some effect on the electronic bell's sound. Anyway, all of this provides tremendous versatility regarding your final decision(s).

So, as you see, this is a rather simple device. It will, however, come in handy when you're involved in telephone projects. As for the power source, this can be any wall style DC transformer rated between 5 and 15 volts. That is the voltage range of the LM555 chip.

Be sure the supply's output doesn't exceed 15 volts. Often, the so-called "battery eliminator" plug-in power units produce voltages higher than advertised. Consequently, a "15-volt" supply might actually be delivering something in the range of 17 volts, and this could fry the 555s.

## CONSTRUCTION

Due to its simplicity, this circuit can easily be constructed with perf-board and point-to-point wiring. However, I fancy printed-circuit boards (PCBs), and have included *Figures 4-2* and *4-3* to facilitate PCB construction. They are, of course, the foil pattern and parts placement, respectively.

Using the PCB, first install the IC sockets and jumpers. The optoisolator only has six pins, so you can either cut two pins off an eight-pin socket, or just use the eight-pin socket as-is. Six-pin sockets are sold for these units if you want to take the trouble to order one.

Next come the resistors, capacitors, diode, transistor and LEDs. Follow that with the relay and metal-oxide varistor. With the board "stuffed" to this point, apply the external voltage, and check for it at pin 5 of U1 and the center contact of the relay. Also, LED1 should illuminate.

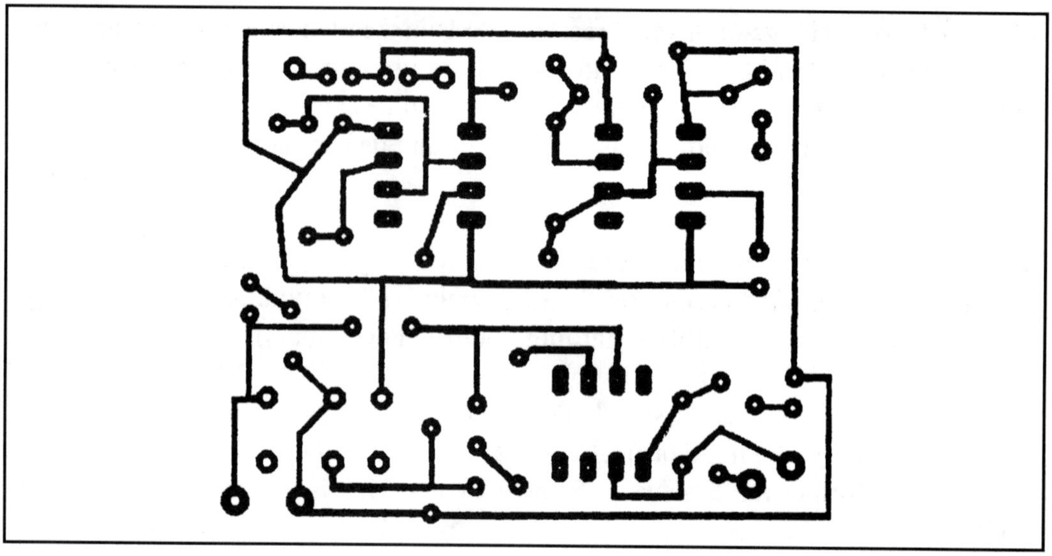

**Figure 4-2.** Electronic bell PCB foil pattern.

If that proves out, disconnect the power and install U1. Now, reconnect the supply and hook an AC voltage between 24 and 48 volts to the "tip" and "ring" points. The relay should activate (you will hear a "click"), and the external supply voltage will now be present at pins 4 and 8 of U2 and U3. This time, both LED1 and LED2 will light.

Assuming everything looks good, disconnect the power, hook up the speaker and install the chips. With the supply reconnected, applying AC to the ring and tip results in the speaker emitting alternating dual tones. Note that LED2 will light each time the AC is applied, indicating a ringer signal.

The last construction step is to permanently connect the tip, ring and power supply leads. Since the wall transformer will probably have some sort of plug on the end of its cable, a matching jack wired to the PCB is the answer for this hookup.

As for the telephone connection, that is best done with an appropriate length of 4-conductor phone cable. One end should have a RJ11 plug. The other end should have the red (ring) and green (tip) leads separated and stripped. The black and yellow wires can be cut off.

Next, solder the tip and ring to the appropriate PCB pads. And, that completes the assembly phase. If you plan to enclose the bell, this would be a fine time to do so.

You will need a case large enough to handle both the board and the speaker. However, those small speaker units designed for desktop use often have sufficient space for the PCB.

## OPERATION

Techniques for using the electronic bell should already be fairly obvious. Once connected to the telephone line, each time the central office rings your number, this de-

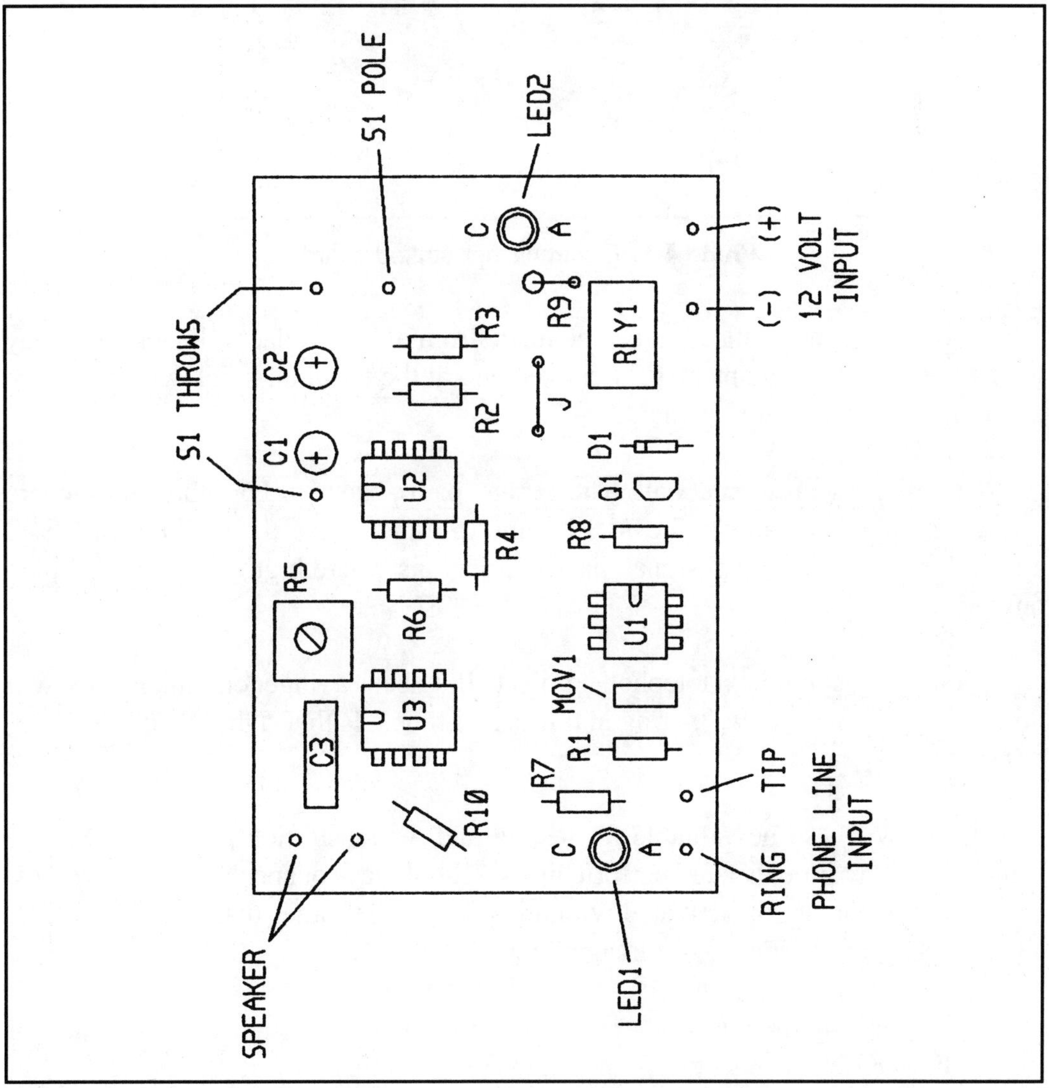


**Figure 4-3.** Electronic bell PCB parts placement.

**Photo 4-1.** Electronic bell stuffed board.

vice also rings. Line connection can be made from any modular wall jack (you may want to use a duplex adapter), or if convenient, at the terminal block where the line comes into your home or business.

As to where this gem is practical? The garage, patio, back yard or other areas where a ringing telephone can't be heard all come to mind. Also, the addition of LED2 allows for a visual "ringer" signal, handy in regions where high noise levels drown out the audio tone.

And, don't forget your next telephone project. If it needs a ringer circuit, this one will do a dandy job. You know, "A ring in time saves nine." Is that right? Well, it's something like that.

The device will also be valuable for testing and demonstration purposes. Or, as a sound effect for "little theater" performances, school plays or special tape recordings. Since the system can be activated with an AC potential other than the telco ringer signal, it has possibilities. Ain't imagination wonderful?

Use yours, and I'm sure many other ideas will come to mind. Such things as practical jokes (BOOOOO!) or other pranks might make use of this project. Just keep it legal, and merciful.

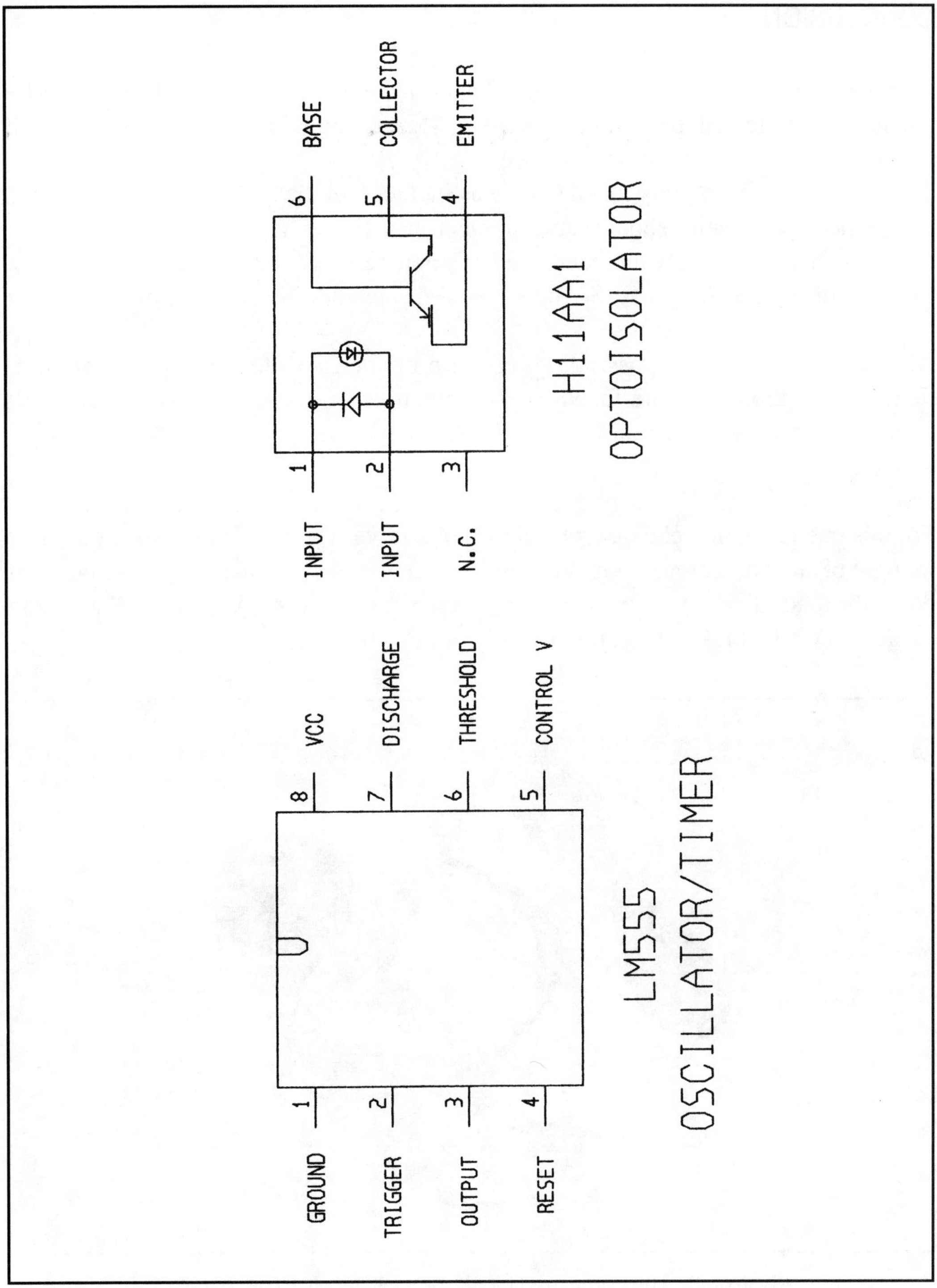


**Figure 4-4.** Electronic bell IC pinouts.

## CONCLUSION

Tote that bail, and ring that bell! What? That isn't right either? Oh well, I try. Maybe you can't tote the bail, but with this project in hand, you will be able to ring the bell.

That, of course, is not why this device was included in the text. Any well-equipped telephone experimenter should have an electronic ringer in his or her arsenal. As a matter of fact, I originally included one in Chapter 2's "Telephone Lab," but frankly, I ran out of room in the case. So, here it is as a separate (but equal) project.

Also, don't overlook the possibilities concerning the first section of this configuration. The telephone line interface/relay combination is used here to activate the oscillators, but that doesn't mean it couldn't be used to energize other "stuff." Think about it.

So, this one is simple, cheap, and fun to build. Give it a try. If you haven't already thought of an application, I can almost guarantee it will come in handy at some point down the road. Thank you, thank you, but save your applause. As they say, "just send money!" (Well, at least tell your friends about this book.)

**Photo 4-2.** The completed electronic bell, ready to be installed in a telephone or separate case.

# CHAPTER 5

# TELEPHONE-ACTIVATED TAPE RECORDER SWITCH

For those of us who enjoy building kits, I try to include one in each book. So, here it is. This is a simple device (at two transistors, three resistors and a capacitor, they don't come much simpler), but one you will find quite handy. It allows you to tape record the traffic on your telephone line, providing a permanent chronicle. Each time the receiver of any telephone on a particular line is lifted, this switch automatically starts a tape recorder. The audio is taken directly from the phone line, so both sides of the conversation are transcribed.

Taping time is limited only by the size of the cassette you place in the recorder, and virtually any machine with auxiliary microphone and remote "on-off" jacks can be used. More will be said about these recorders in the Operation section.

All in all, you will find this automatic tape recorder switch an admirable addition to your telephone system. It will furnish a complete record of all telephone activity, and that can be very beneficial for recounting details of business and personal phone conversations. One warning, however. In some areas, use of this device may be illegal. So, it's best to check with your local authorities. It is my understanding, though, that in most jurisdictions if one of the parties involved is aware of the recording, said recording is perfectly legal.

## THEORY

From what we learned in Chapter 1, the basic principle behind this device should be somewhat obvious. Remember, when you pick up the telephone's receiver, the line voltage drops from around 48 volts to about 5 volts. Hence, what we need here is a dedicated switch that detects that voltage drop and activates the tape recording unit.

Additionally, we need to send the on-line signal to the microphone input of the tape machine. Then, whatever is being said or sent over the line will be recorded.

All this sounds simple enough, and in reality, it is. As with many electronic challenges, there are a number of different ways to approach it. Two possible solutions might be the old standby "window comparator" or a regulation style circuit that triggers a transistor switch.

While those and many others will work, the kit used here provides another clever design to solve the task. Designated the "TEL-SW-1," this kit comes to us from The Electronic Rainbow, and is distributed by them and a number of mail-order companies. See parts and source lists for information on obtaining the TEL-SW-1.

So, let's take a look at how The Electronic Rainbow approaches this project. I think you will appreciate the simplicity employed in their design. As I've often said in the past, just because it's electronic doesn't mean it has to be difficult.

Rainbow design engineer Paul Turner has devised a simple method to detect the voltage change. Referring to *Figure 5-1*, the incoming telephone line is first fed to a voltage divider consisting of resistors R1 and R2. This pair sets a voltage that causes transistor Q1 to conduct.

That action keeps transistor Q2 from conducting, thus preventing the tape recorder from operating. However, when the line battery voltage drops due to a telephone receiver being lifted, the voltage at Q1's gate is too low for it to conduct. This allows Q2 to conduct, which triggers the tape recorder.

The audio signal on the line is introduced to the tape machine via an appropriate phone jack. The point marked "line/common" handles the negative or "ground" side, while the signal at "line" is capacitively coupled to the positive recorder input.

Since various tape recorders use different "grounding" schemes, a third connection point, "audio common," is provided to handle a "hot shield" scenario. Rainbow provides the required capacitor for this type of hookup, and more will be said regarding this in the next section.

Well, there it is! Simple, isn't it? Simple, but very effective. OK, let's move on to building this device.

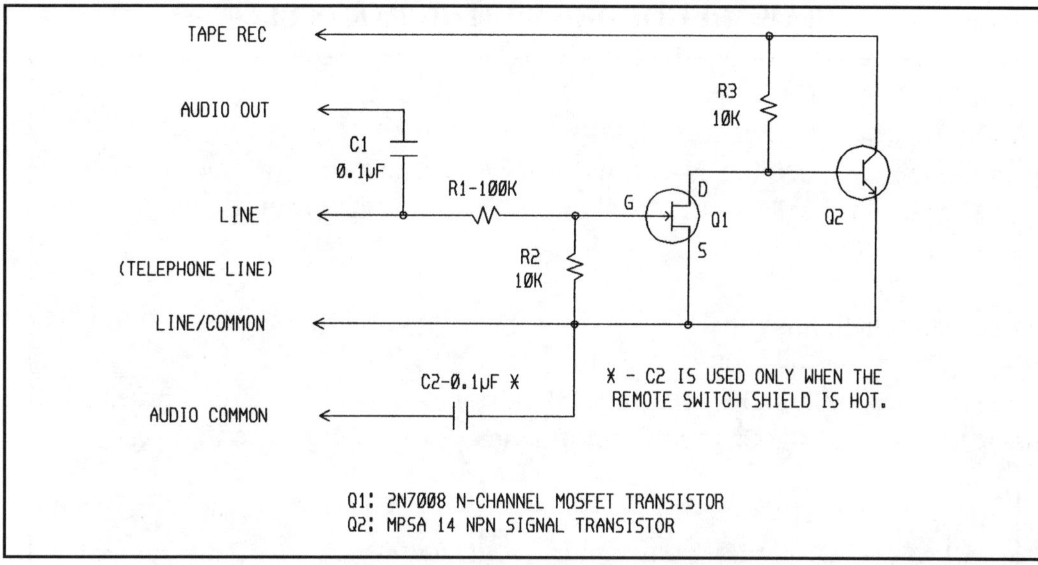


**Figure 5-1.** Tape recorder switch schematic.

## CONSTRUCTION

As can be seen in *Photo 5-1*, this is a very small device. The printed-circuit board (PCB) measures just 9/16 of an inch (14 mm) by 1 1/16 inches (27 mm), but contains all components. The kit, of course, comes with a PCB; however, if you would like to make your own, *Figure 5-2* illustrates the foil pattern.

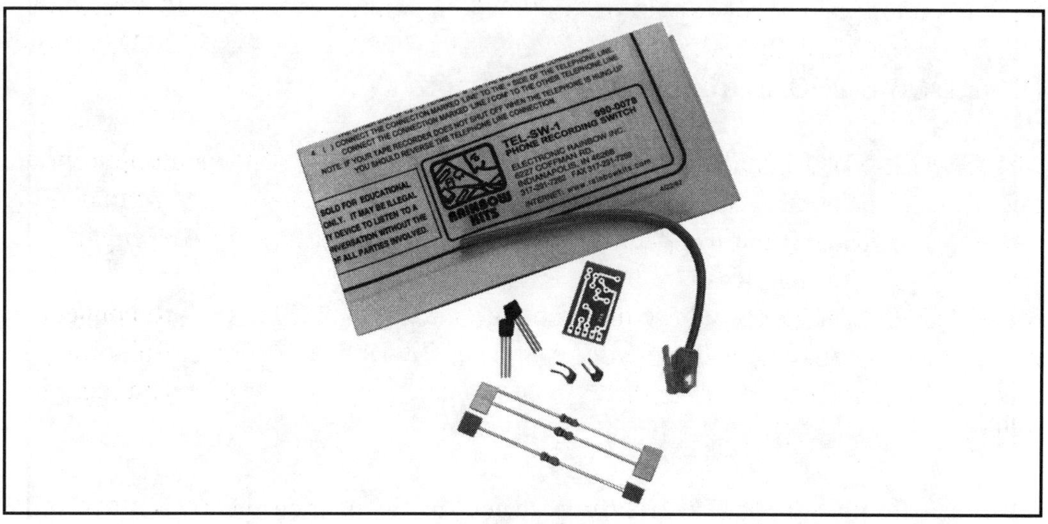


**Photo 5-1.** The complete Rainbow kit TEL-SW-1 as it comes out of the package.

## TAPE RECORDER SWITCH PARTS LIST

SEMICONDUCTORS:

| | |
|---|---|
| Q1 | 2N7008 N-Channel MOSFET Transistor (Jameco) |
| Q2 | MPSA14 NPN Signal Transistor (Digi-Key) |

RESISTORS:

| | |
|---|---|
| R1 | 100,000 Ohm 1/4W Resistor |
| R2, 3 | 10,000 Ohm 1/4W Resistors |

CAPACITORS:

| | |
|---|---|
| C1, 2 | 0.1 μF Monolithic Capacitors |

MISCELLANEOUS:

PCB Materials or Board, Solder, Hookup Wire, Modular Telephone Cable w/RJ11 Plug, Enclosure (if desired), Etc.

TEL-SW-1 PHONE RECORDER SWITCH KIT:

The TEL-SW-1 kit can be purchased from the following sources:

Electronic Rainbow, Inc.
6227 Coffman Road
Indianapolis, IN 46268
Phone: (317) 291-7262

Tech America
P.O. Box 1981
Fort Worth, TX 76101
Phone: (800) 877-0072

Part #: 990-0078

It usually runs about $12.95, and these are just two sources for the kit. Many companies that carry Rainbow Kits will have this one available.

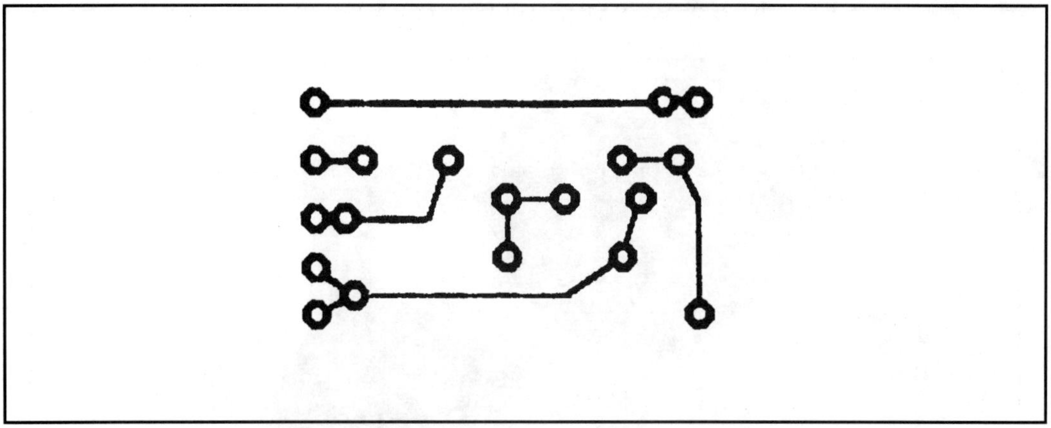

**Figure 5-2.** Tape recorder switch PCB foil pattern.

Using the parts placement diagram, *Figure 5-3*, as a guide, first install resistors R1, R2 and R3, then capacitor C1. Next come transistors Q1 and Q2. Regarding these transistors, both have the same shape (TO-92), so be careful to get the right component in the right place. As for orientation, the flat sides face toward the center of the board.

With that done, it's time to connect the cables. The short piece of telephone cable that comes with the kit is a little inconvenient, so I whacked a modular plug off one end of a six-foot extension line and connected the remainder of the assembly to the TEL-SW-1 board. The green (tip) goes to "LINE" and the red (ring) to "LINE/COMMON."

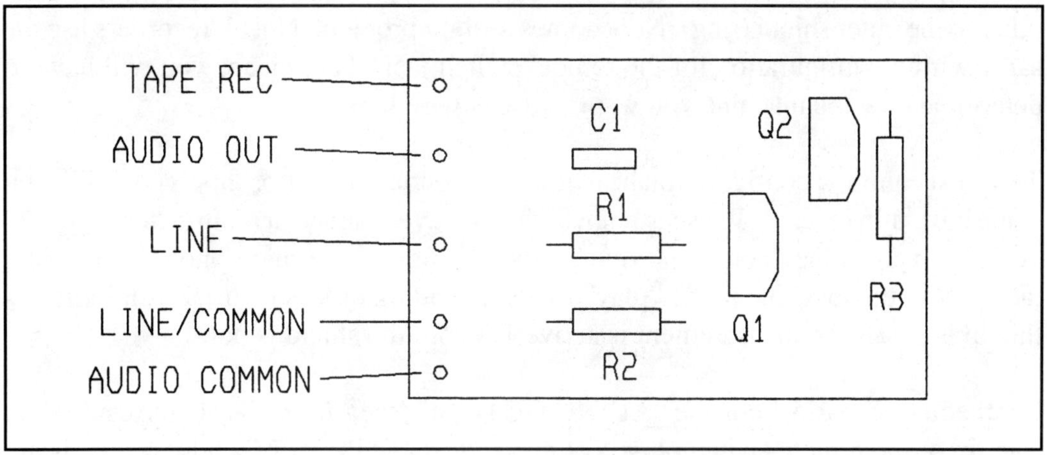


**Figure 5-3.** Parts placement.

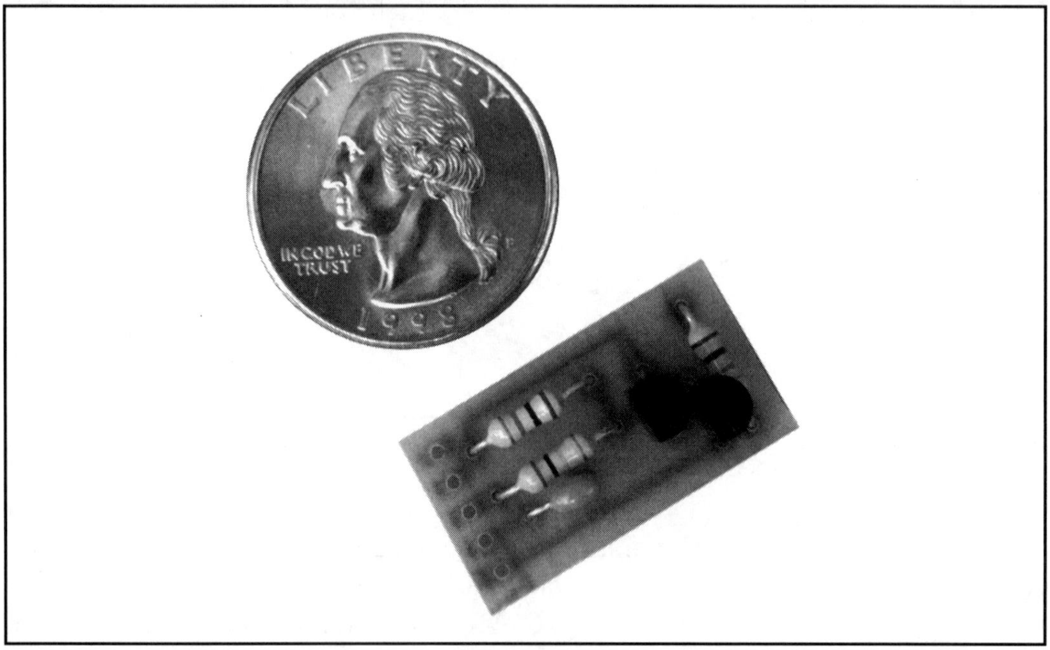


**Photo 5-2.** The stuffed board of the TEL-SW-1 phone recording switch. Note the small size compared to the quarter.

For the recorder's microphone and switch jacks, 36-inch (915 mm) lengths of dual-conductor speaker wire work nicely. The remote "on-off" jack requires a small 3/32-inch phone plug, while the mic uses the standard 1/8-inch receptacle.

For each, one conductor of the speaker wire is soldered to the center lug (tip), and the other to the outer shield (ring). Now comes a slight problem. Not all recorders use the same wiring configuration for the remote switch jack. That means you will have to determine how the machine you want to use is wired.

The most common configuration has the center conductor, or tip, "hot" (positive), but some have it in reverse. To ascertain which you have, connect a multimeter set for 20 volts DC across the "common" (outer shield) rings of the microphone and switch jacks. With the machine in the "play" mode, a reading of less than 1/2 volt indicates the tip is hot. If the measurement is above 1/2 volt, the shield is hot.

For the first scenario, hook the "AUDIO OUT" line (see *Figure 5-4*) to the center lug. The tip of the remote switch plug is connected to "TAPE REC" and the shields are both hooked to the "LINE/COMMON" pad.

For the second situation, the tip of the remote switch goes to "LINE/COMMON", and the shield goes to "TAPE REC." For the mic plug, the tip connects to "AUDIO OUT," and the shield to "AUDIO COMMON" through the included 0.1-microfarad capacitor. *Figure 5-4* furnishes complete details.

Once the wiring is completed, the unit is ready to be installed in your telephone system. If desired, a small plastic box can be used to house and protect the device. Also, due to its tiny size, the unit easily fits inside many telephones.

## TEST AND OPERATION

Testing is easy. With the TEL-SW-1 plugged into both the recorder and the modular wall jacks, lift the telephone receiver and the tape should start moving. This is presuming you have the machine in the "record" mode. When the phone is hung up, the tape will stop. If it doesn't, reverse the "line" connections.

If nothing at all happens, double check your parts placement and wiring. With boards this small, solder bridges are quite common. Also, since the device is powered by the phone line itself, make sure the modular plug is securely inserted in the modular wall jack.

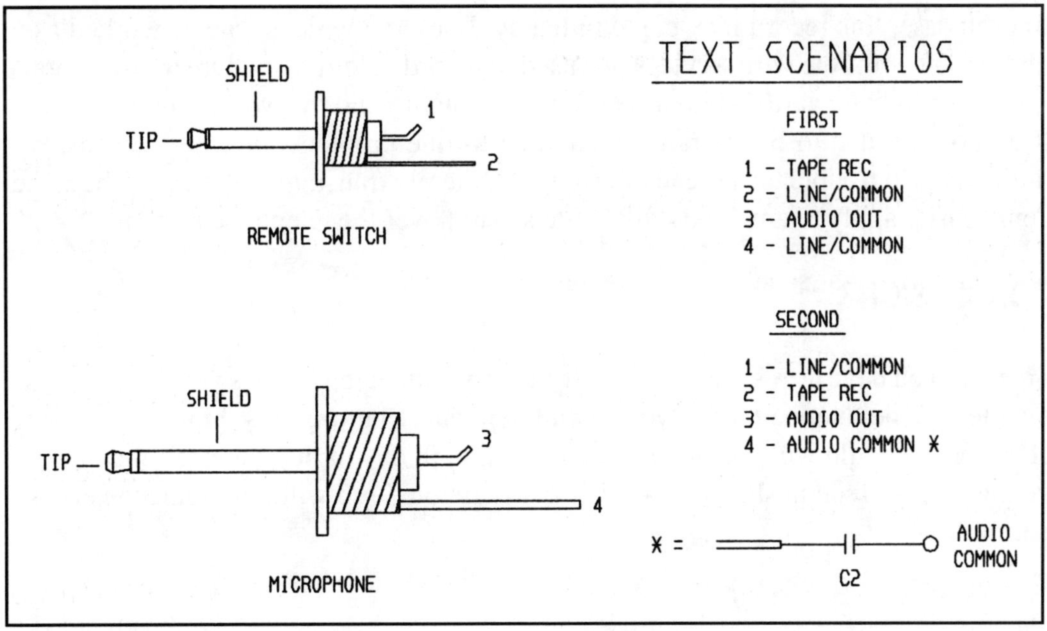


**Figure 5-4.** Tape recorder hookup.

As promised, let's talk a little more about the type of tape recorder you will need. What we want here is a machine that has jacks for both an auxiliary microphone and remote power switch. These aren't as common as they were, say, ten years ago, but can still be found at retail electronics and discount stores.

They usually come in packages 8 to 10 inches long by 5 to 6 inches wide, thus are larger than many cassette recorders. They also usually include a microphone with twin phone plugs, one for the mic and the other for the built-in slide switch. The switch, of course, is used to halt (pause) the recorder, and is handy when dictating information.

In every case I've encountered, the "switch" plug is the smaller of the two. However, with electronics, nothing is etched in stone, so it is advisable to confirm which plug is which. You probably won't hurt anything if you get them backwards, but the device and/or recorder will not work properly.

OK, with everything set up, put the tape machine in the "record" mode. Now, whenever a phone on the line is picked up, the conversation or other information will be placed on magnetic tape. Other information could include faxes, telexes or similar data transfers, usually from computer modems.

In each case, this record may come in handy. Back in the days when I worked as a photographer for the wire services, we used to record incoming picture transmissions (photofaxes) from various bureaus around the country and the world. In this fashion, if a photo needed to be "retransmitted" due to line problems or other reasons, we didn't have to trouble the bureau of origin. We merely transmitted the tape. The same could easily apply to any kind of data transferred over the telephone system.

## CONCLUSION

So, there you have it. A super-duper-nifty-thrifty gadget that allows you to "tap" your own telephone. Notice I said "your own telephone." Putting this thing on someone else's phone could win you an extended vacation in one of our country's many marvelous state or federal slammers — especially if you do so without "someone else's" permission.

But, I know you wouldn't do something like that! So the preceding is certainly an unneeded cautionary declaration. All kidding aside (Ha, Ha), this device will prove to be most useful. I'm certain you will be surprised at how often a taped record of personal conversations and/or business calls will be beneficial.

For example, you may need to double-check addresses or telephone numbers given to you over the phone, confirm directions or instructions, or verify other details of conversations. Additionally, in the unfortunate circumstance that you should receive harassing or pornographic phone calls, it's nice to have those calls (for the authorities, of course) on tape.

Anyway, give this one a try. It's easy to build, inexpensive, and results in a highly pragmatic device. If nothing else, the next time that obstinate caller (we all know one of those) insists he or she told you "Da Da Da" during your last phone encounter, you can whip out your trusty tape and show them it was really "De De De" — so there!

# CHAPTER 6

# A DESK PHONE AND AUXILIARY SPEAKERPHONE

Originally, this chapter was going to cover a standard "touch-tone" desk style telephone. But, as I got into that subject I began to realize it was, well, pretty boring. Not terribly boring, just pretty boring. So, I decided to add some "spice."

The spice comes in the form of cinnamon. No, not really! More to the point, it comes as a "two-part" chapter. The first part will briefly cover the original goal, while the second part addresses a more interesting project — a speakerphone!

Through an arduous and exhausting search I was able to locate an integrated circuit that made this task feasible. Well, actually, JDR Microdevices sent me a new catalog carrying the Motorola MC 34018P Speakerphone chip. But, I found it and this chapter is primarily the result of that discovery.

All right now, don't get ugly! I think you will like this project. In addition to being fun, it will also prove to be very useful. I mean, you never know when you might need a speakerphone, do you? OK, so maybe you do know. We can debate that later.

## PART ONE: REGULAR OL' PHONE

As promised, let's look at the basics of the familiar "desktop" pushbutton telephone. First, *Figure 6-1* illustrates the anatomy (electrically speaking) of the rotary dial pulse telephone that was a staple of the industry for so many years. As can be seen, it wasn't exactly state of the art. Well, in the beginning it was, but by the time they were fazed out, these "dinosaurs" had seen their day.

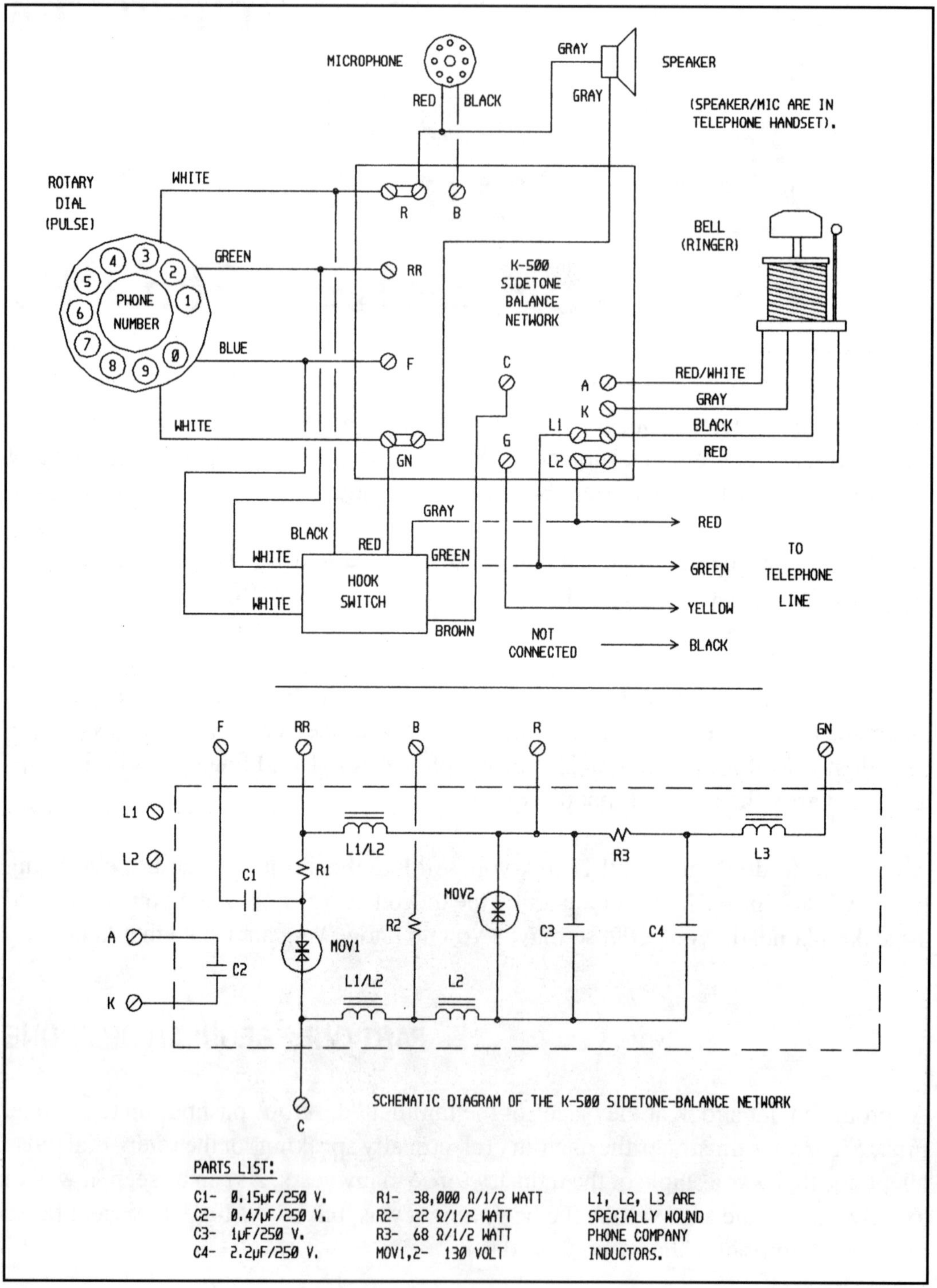


**Figure 6-1.** Rotary dial (pulse) telephone circuit.

**Photo 6-1.** A typical Bell K-500 sidetone-balanced network. This view shows the interior of the normally enclosed module.

Anyway, they consisted of an electric bell, a hook switch, a rotary dial, a handset with speaker and microphone, and what "Ma Bell" affectionately called a "sidetone-balance network" (SBN). For our desk phone, we are going to throw out the electric bell and rotary dial, and keep the rest. Note that *Figure 6-1* also shows the internal configuration of the normally sealed sidetone net.

In place of the dialing mechanism, the desk phone will utilize a dual-tone multifrequency (DTMF) tone generator, and the bell, if you want one, will be replaced with an electronic version. The sidetone effect and telephone line (tip/ring) interface will be handled by the sidetone-balance network.

**Photo 6-2.** Interior view of the older style rotary-dial telephone.

As for the handset, that will be the same. The hook switch will be essentially identical in principle, but a much simpler switching arrangement. Our unit will need only a normally closed (NC) pushbutton switch instead of the elaborate device found in the mechanical/electric units.

## THEORY

*Figure 6-2* is the schematic diagram of our desk telephone, and I chose the Texas Instruments TCM5089 Tone Encoder as the generator chip. In fact, this circuit is basically the TI recommended application of the TCM5089 chip. This IC was selected because of its simplicity, low cost and availability.

As seen, the circuit is fairly simple and straightforward. The SBN manages interfacing for both the handset and the telephone line, while the 5089 circuitry provides "dialing." Incidentally, using the sidetone network (officially called the K-500 sidetone-balance network) does require the "carbon" microphone and 25-ohm speaker found in the older style phones.

Notice the K-500 network screw terminals are designated "A, B, C, F, K, R, GN, L1, L2 and RR." Each of these connects to a section of the overall circuit, as is detailed in *Figure 6-2*.

Regarding the tone generator electronics, the 5089 produces the DTMF tone pairs and is crystal controlled. As you have probably noticed by now, the use of a crystal for accuracy is common among telephone encoders and decoders.

Diodes D1 through D4 are arranged as a bridge to protect the circuit against reverse line voltages. Diode D5 is a 1N4743 zener that regulates the line voltage to a maximum of 13 volts. This, of course, prevents damage and/or destruction of the 5089.

Transistors Q1 and Q2, as well as resistors R1 through R4, comprise a switching arrangement that inhibits the microphone during dialing. This is done to prevent ambient sound from entering the mic and possibly interfering with the dial tones.

Capacitor C1 is employed for both bypass and high-frequency suppression duties. And, last, but certainly not least, is the keypad, that allows entry of the desired phone number. This is the standard "X-Y matrix" design that shorts one row to one column.

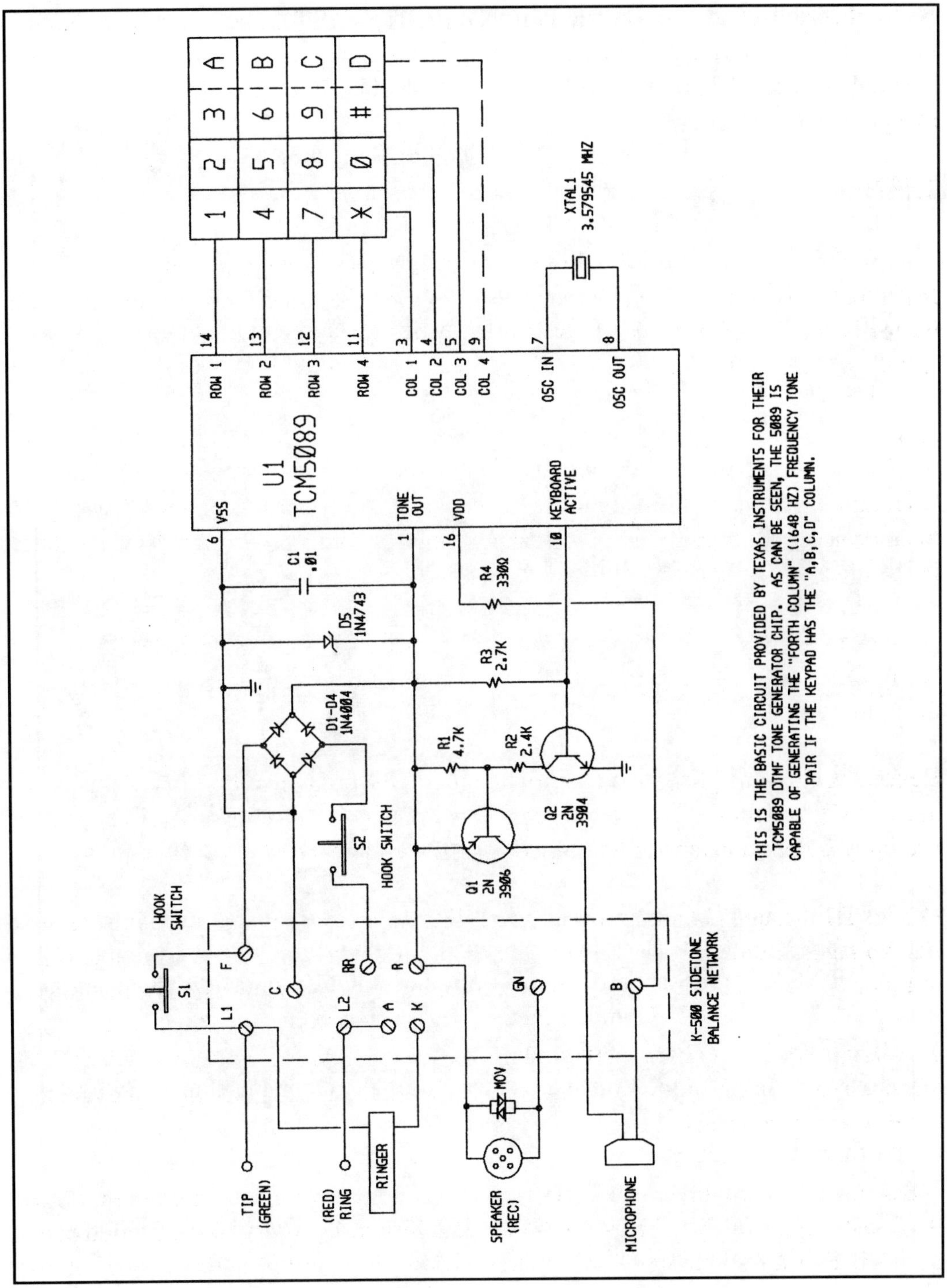


**Figure 6-2.** TCM5089-based dual-tone desk phone.

## DESK PHONE PARTS LIST

SEMICONDUCTORS:

| | |
|---|---|
| U1 | TCM5089 Tone Encoder (JDR, Jameco) |
| Q1 | 2N3906 PNP Signal Transistor |
| Q2 | 2N3904 NPN Signal Transistor |
| D1-4 | 1N4004 400V Rectifier Diodes |
| D5 | 1N4743 13V Zener Diode |
| MOV | 130V Metal-Oxide Varistor (May be in the handset) |

RESISTORS:

| | |
|---|---|
| R1 | 4,700 Ohm 1/4W Resistor |
| R2 | 2,400 Ohm 1/4W Resistor |
| R3 | 2,700 Ohm 1/4W Resistor |
| R4 | 330 Ohm 1/4W Resistor |

CAPACITORS:

| | |
|---|---|
| C1 | 0.01 μF Disk/Mylar Capacitor |

OTHER COMPONENTS:

| | |
|---|---|
| K-500 | K-500 Type Bell Sidetone-Balance Network (Old phone or local Bell Co.) |
| XTAL1 | 3.579545 MHz Crystal |
| KEYBRD | 12/16-Key X-Y Matrix Keyboard |
| HANDSET | Standard Telephone Handset w/Carbon Microphone and 25 Ohm Magnetic Speaker (receiver) (see text) |
| RINGER | Several Options - see text |
| HKS | DPST Spring Loaded, Normally Open Pushbutton Switch |

MISCELLANEOUS:

Solder, PCB Materials or Perf-Board, Hookup Wire, Old Style Telephone, Hardware, IC Socket, Etc.

**Photo 6-3.** Interior view of an older pulse-dialing "Ma Bell" unit. Note the electric bell and sidetone-balanced network.

Both the handset and the sidetone network can be borrowed (stolen) from an older desk phone and the rest of the components are easily obtained from a variety of suppliers. Another reason for using the TCM5089 chip is that it is just a dual-tone generator. Unlike some of the other available encoder chips, the 5089 doesn't have mute and memory features, which makes it less complex to work with.

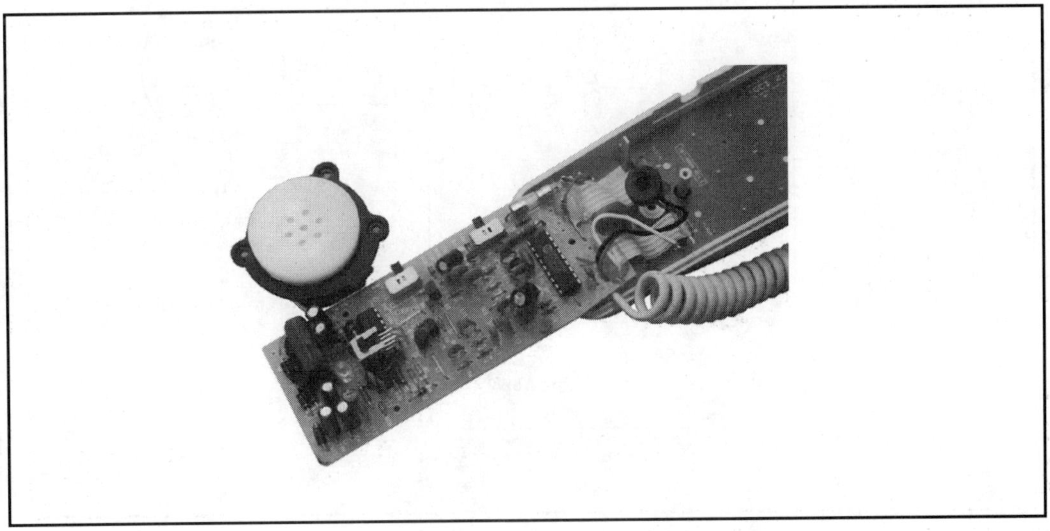

**Photo 6-4.** Internal view of modern self-contained electronic telephone.

## CONSTRUCTION

As for construction technique, this circuit is easily arranged on perf-board, then placed inside an old telephone case. Or, it could be incorporated into any type of "specialized" equipment that requires telephone line access. A printed-circuit board (PCB) can also be used but is by no means necessary with this system.

If you wish to include a "ringer" device, the Electronic Bell in Chapter 4 would do right nicely. However, if you are looking for a simpler design, I'm including the circuit in *Figure 6-3*. Here, a Motorola MC34017P integrated circuit is employed to provide the familiar "warble" tone so often associated with today's electronic telephones.

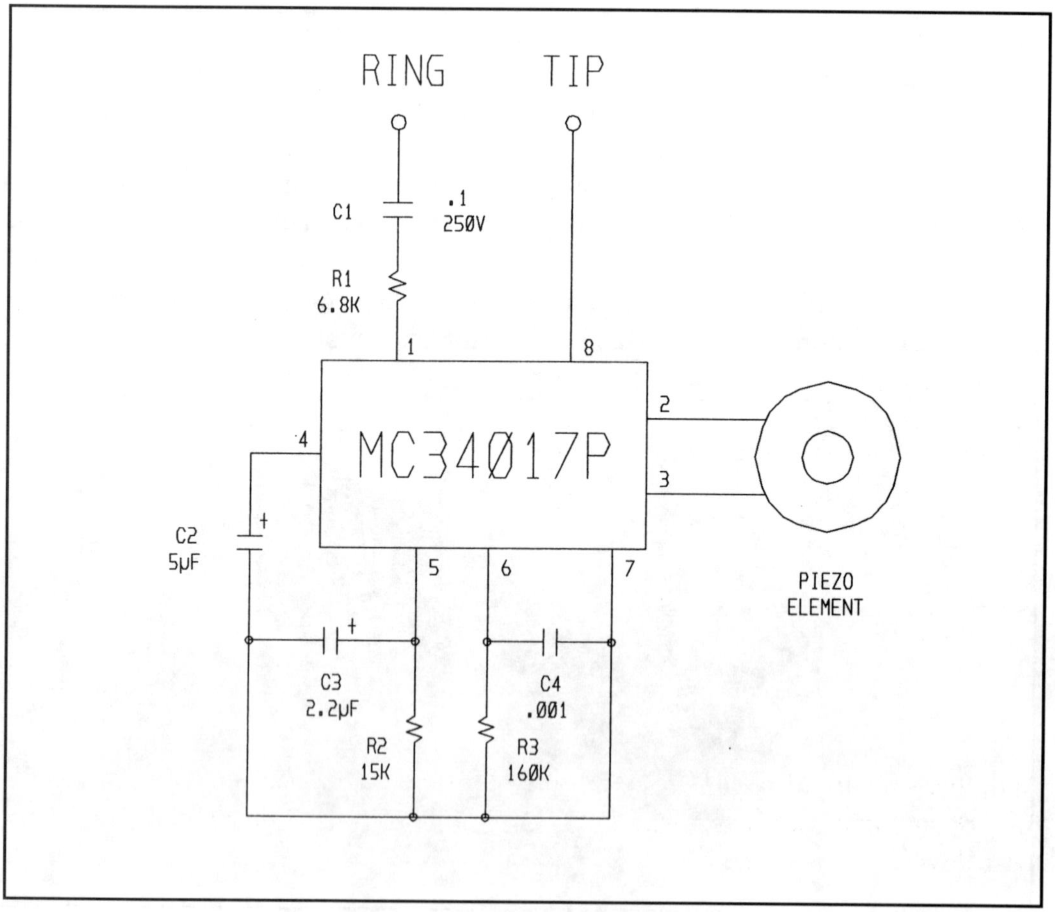


**Figure 6-3.** MC34017 ringer circuit.

## MC34017 RINGER CIRCUIT PARTS LIST

SEMICONDUCTORS:

| | |
|---|---|
| U1 | MC34017P Ringer Detector/Tone Generator |
| PIEZO | Piezo Sounding Element (Buzzer) |

RESISTORS:

| | |
|---|---|
| R1 | 6,800 Ohm 1/4W Resistor |
| R2 | 15,000 Ohm 1/4W Resistor |
| R3 | 160,000 Ohm 1/4W Resistor |

CAPACITORS:

| | |
|---|---|
| C1 | 0.1 µF 250V Mylar Capacitor |
| C2 | 5 µF Electrolytic Capacitor |
| C3 | 2.2 µF Electrolytic Capacitor |
| C4 | 0.001 µF Disk/Mylar Capacitor |

MISCELLANEOUS:

Solder, PCB Materials or Perf-Board, IC Socket, Hookup Wire, Hardware, Case (If Desired), Etc.

## CONCLUSION

And, that's about all there is to it. Once completed, this phone should work beautifully on any line capable of handling DTMF dialing — that's just about all of them these days.

## PART TWO: AUXILIARY SPEAKERPHONE

**Photo 6-5.** Our auxiliary speakerphone.

OK, so much for the "simple" part of this chapter. This next section gets a little more complicated, but it produces a far more versatile device.

If you have never used a speakerphone, this project will provide a treat. They have come a long way in terms of sound quality, sensitivity and ease of use over the last few years. A far cry from the original "Bell" devices.

Using the Motorola MC34018P, four transistors and a double handful of discrete components, the finished product is a "single chip" full-function speakerphone. The 34018 provides a system with remarkable sensitivity, background noise suppression and all-around high sound quality.

Adding a Motorola MC34017P Ring Detector/Tone Generator and a few discrete components will result in an almost "standalone" phone. I say "almost" because you don't have a keypad for dialing, but you will hear the ringer when a call comes in.

All right, enough tooting the 34018's horn. Let's get down to the "nuts and bolts" of this project. After all, that what we're here for, right? Right!

## THEORY

In all, the MC34018P is one he...no, I can't use that word here...let's say one *heck* of an integrated circuit. In a single 28-pin package, you will find a microphone amplifier, power amplifier (speaker), transmit and receive attenuators, background sound level monitoring, and regulated voltages for both internal and external circuits.

All of this provides the necessary components for a complete speakerphone. And, if you want to get really fancy, the 34018 can be combined with other chips, such as the Motorola MC34014P Speech Network and/or many of the available tone encoders, to create a complete "speaker/handset" system.

With that said, let's take a gander at our speakerphone. *Figure 6-4* illustrates the schematic of a design that is the basic Motorola suggested layout. I have made a few component value changes, and added the ring detection circuit and "ON" LED, but otherwise it is the circuit supplied in the data sheets.

The 34018, and associated parts, is the heart of this configuration, controlling both the microphone and speaker. External resistors and/or capacitors handle such tasks as transmit-receiver attenuation, output levels and line impedance matching.

Since all speakerphones employ a "half-duplex" operation, that too is accomplished by the IC. The purpose of half-duplex (either the microphone or speaker active, but not both at any given time) is to prevent acoustic feedback due to the close proximity of the mic and speaker. Not a desirable effect — feedback, that is.

Background level monitoring furnishes a stronger and far clearer signal to the phone at the other end. If you've ever had the experience of talking with someone who was using one of the early speakerphones, you well remember that "hollow" sound characteristic of those devices. The 34018 background monitoring feature virtually eliminates that problem by balancing the "noise" (ambient sound) with the voice signal.

Switch S2, a single-pole/single-throw (SPST) unit, allows you to disable the microphone for "privacy" listening. This can be a handy feature if you don't need to participate in a conversation. Finally, potentiometer R28 allows adjustment of the speaker output volume. R28 also affects the speed of the switching between transmit to receive and back, but frankly, I didn't notice much difference in that function regardless of where the potentiometer was set.

The top half of the diagram depicts the telephone line interface circuit needed to connect the 34018 to the line. This is basically amplification and matching, with the following representing a few of the highlights.

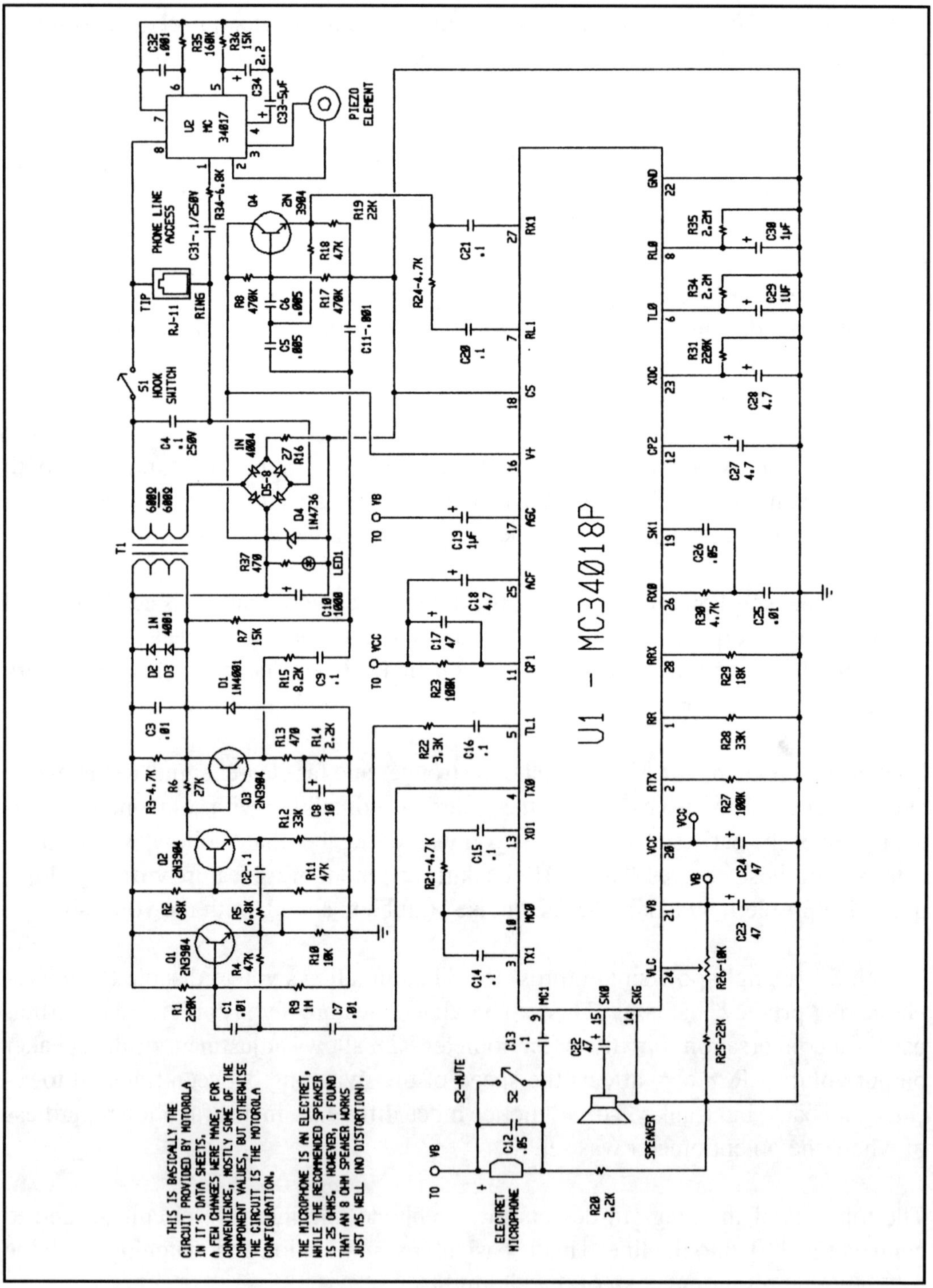


**Figure 6-4.** MC34018 speakerphone circuit.

Transformer T1 is a standard 600-ohm-to-600-ohm line matching unit that provides the proper circuit impedance. Diode D4 regulates the voltage to a safe 6.8 volts that prevents the higher line voltage from damaging U1. And, diodes D5 through D8 form the familiar bridge rectifier that protects against line voltage polarity reversal.

The single-pole/single-throw (SPST) switch S1 is the hook switch which merely disconnects the "tip" side of the incoming line, thus effectively taking the speakerphone "off-line." Last, U2 and friends comprise a "ring detection" system that lets you know someone is calling. This is the same circuit as described in Part One of this chapter.

By the way, the 1,000-microfarad capacitor (C10) in the power section is essential. I tried several smaller values, and the results were not of the healthy kind. They ranged from an "oscillating" speaker emission to a high squeal reminiscent of feedback. Hence, keep that capacitor at 1,000 or above.

## SPEAKERPHONE PARTS LIST

| SEMICONDUCTORS: | |
|---|---|
| U1 | MC34018P Speakerphone (JDR, Motorola) |
| U2 | MC34017P Ring Detector/Tone Generator (Motorola) |
| Q1-4 | 2N3904 NPN Signal Transistors |
| D1-3, 5-8 | 1N4004 Rectifier Diodes |
| D4 | 1N4736 6.8V Zener Diode |
| LED1 | T-1 3/4 Jumbo Light-Emitting Diode (Any Color) |
| PIEZO | Piezo Sounding Element (Buzzer) |
| RESISTORS: | |
| R1, 31 | 220,000 Ohm 1/4W Resistors |
| R2 | 68,000 Ohm 1/4W Resistor |
| R3, 21, 24, 30 | 4,700 Ohm 1/4W Resistors |
| R4, 11, 18 | 47,000 Ohm 1/4W Resistors |
| R5, 34 | 6,800 Ohm 1/4W Resistors |
| R6 | 27,000 Ohm 1/4W Resistor |
| R7, 36 | 15,000 Ohm 1/4W Resistors |

(Continued Next Page)

| | |
|---|---|
| R8, 17 | 470,000 Ohm 1/4W Resistors |
| R9 | 1,000,000 Ohm 1/4W Resistor |
| R10 | 10,000 Ohm 1/4W Resistor |
| R12, 28 | 33,000 Ohm 1/4W Resistors |
| R13, 37 | 470 Ohm 1/4W Resistors |
| R14, 20 | 2,200 Ohm 1/4W Resistors |
| R15 | 8,200 Ohm 1/4W Resistor |
| R16 | 27 Ohm 1/4W Resistor |
| R19, 25 | 22,000 Ohm 1/4W Resistors |
| R22 | 3,300 Ohm 1/4W Resistor |
| R23, 27 | 100,000 Ohm 1/4W Resistors |
| R26 | 10,000 PCB Potentiometer |
| R29 | 18,000 Ohm 1/4W Resistor |
| R32, 33 | 2,200,000 Ohm 1/4W Resistors |
| R35 | 160,000 Ohm 1/4W Resistor |

CAPACITORS:

| | |
|---|---|
| C1, 3, 7, 25 | 0.01 μF Disk Capacitors |
| C2, 9 | 0.1 μF Disk Capacitors |
| C4,13-16, 20, 21, 31 | 0.1 μF 250V Mylar Capacitors |
| C5, 6 | 0.005 μF Disk Capacitors |
| C8 | 10 μF Electrolytic Capacitor |
| C10 | 1,000 μF Electrolytic Capacitor |
| C11, 32 | 0.001 μF Disk Capacitors |
| C12, 26 | 0.05 μF Disk Capacitors |
| C17, 22-24 | 47 μF Electrolytic Capacitors |
| C18, 27, 28 | 4.7 μF Electrolytic Capacitors |
| C19, 29, 30 | 1 μF Electrolytic Capacitors |
| C33 | 5 μF Electrolytic Capacitor |
| C34 | 2.2 μF Electrolytic Capacitor |

OTHER COMPONENTS:

| | |
|---|---|
| MIC | Electret Microphone (2-Conductor) |

(Continued Next Page)

| | |
|---|---|
| SPKR | 8 Ohm Speaker (Any Size) |
| S1, 2 | SPST Slide Switches |
| T1 | 600-to-600 Ohm Matching Transformer (Jameco #: 141621) |

MISCELLANEOUS:

PCB Materials, Solder, Hookup Wire, IC Sockets, Case, Modular Telephone Cable w/RJ-11 Plug, Hardware, Dry Transfer Lettering, LED Holder, Etc.

## CONSTRUCTION

Since I had a case with a nice big speaker already installed, and a perfect slot for the volume control, I used that to house this gem. I think this enclosure began life as a toy electronic drum, but I'm not sure. I robbed it from something.

Anyway, the case now dictated the PCB size, which turned out to be 5 inches (127 mm) by 3 1/2 inches (89mm). I also had to cut a small section out of both left PCB corners to accommodate mounting posts, but that still left just enough room to install the rather substantial number of components. This board ends up very "tight," so watch your soldering (assuming you stick with the included PCB foil pattern).

*Figures 6-5* and *6-6* are included to help you construct this project. They are the PCB "foil" or solder-side pattern and parts placement diagram, respectively.

The first step is to place all the jumpers. Several of these go under, or at least start under, U1's IC socket, so these need to go on first. Next come the two sockets, then the discrete components.

Now install the transformer, potentiometer, diodes and transistors. That leaves only the lead wires for the switches, microphone, speaker, piezo and "ON" light-emitting diode (LED). Using appropriate lengths of hookup wire, first connect them to the PCB, then to the individual components.

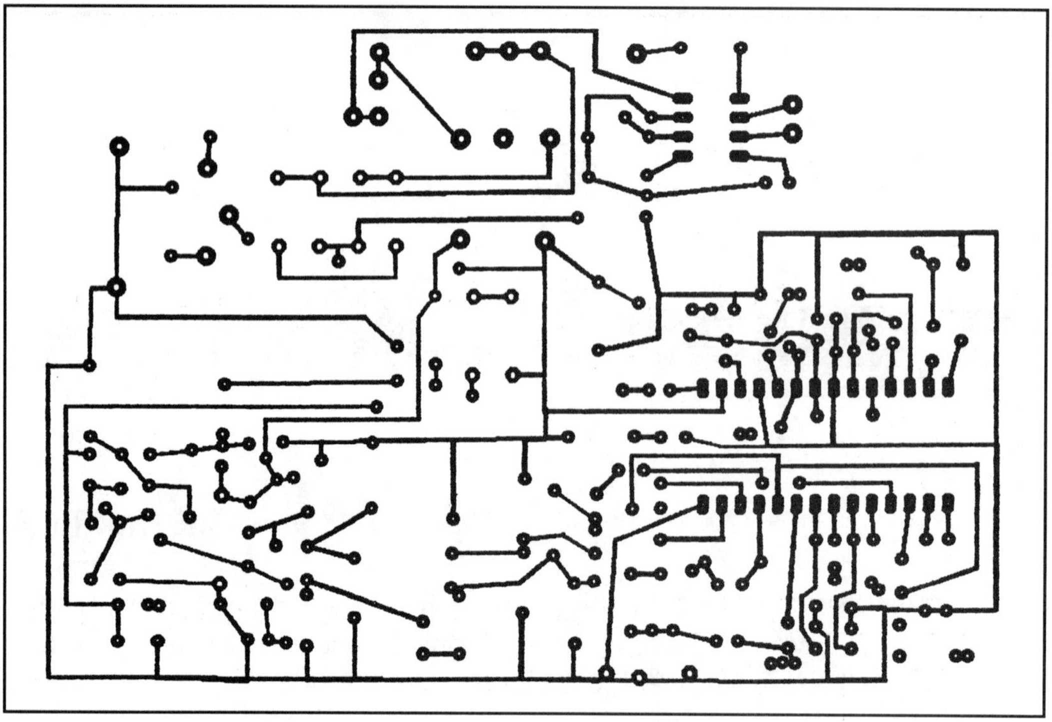

**Figure 6-5.** Auxiliary speakerphone PCB foil pattern.

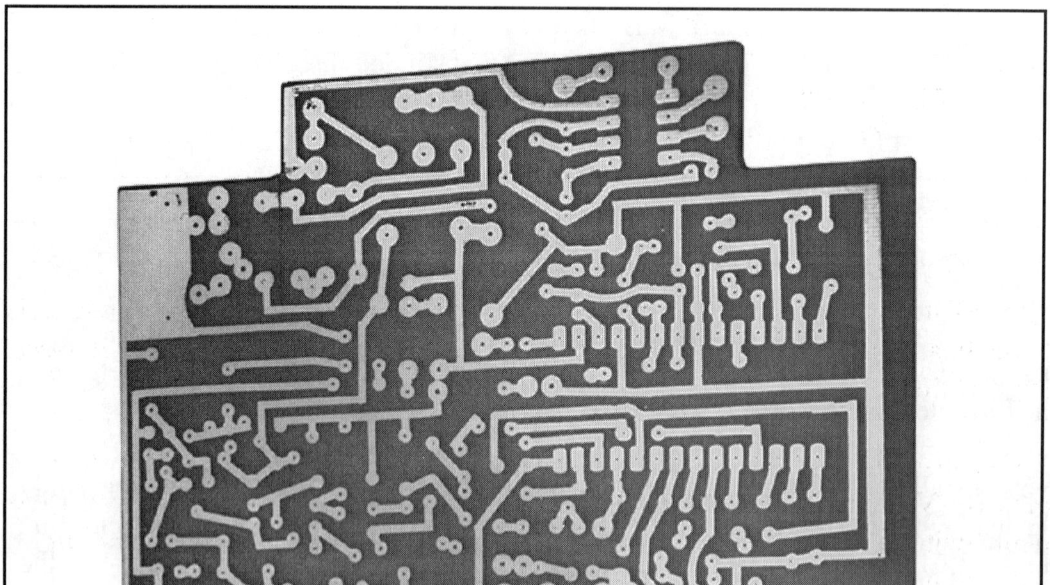

**Photo 6-6.** Note the tight PCB layout.

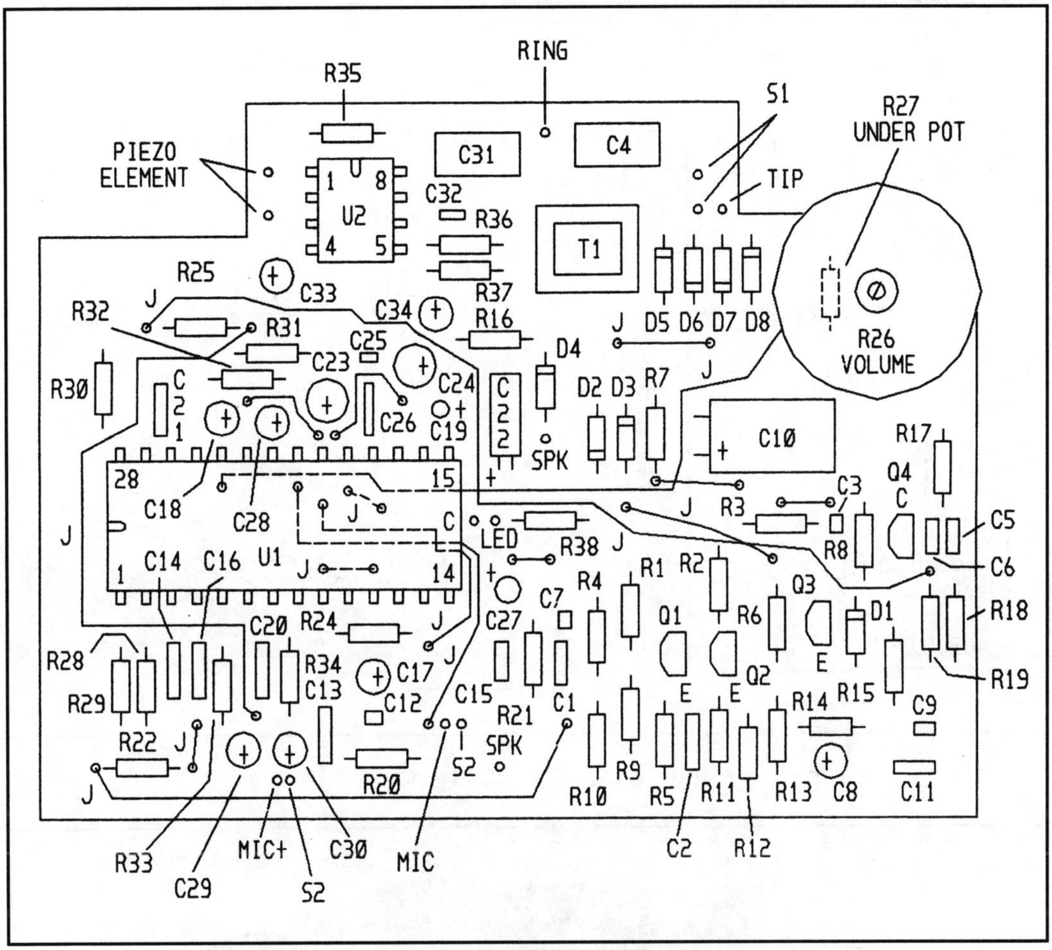


**Figure 6-6.** Speakerphone parts placement.

The last step is to hook up a length of modular cable, with an RJ11 plug on one end, to the board pads marked "tip" and "ring." The red wire is the "ring" and the green lead is "tip."

Before installing the ICs, plug the cable into a telephone jack and check the voltage at pins 16 (V+) and 22 (GND) of U1. This should read about 6.8 volts whenever the hook switch (S2) is in the "off-hook" position. If all is well, the two chips can be plugged into their sockets.

With that done, set the board aside, and turn your attention to the case you have selected. Drill/cut whatever openings needed for the various externally accessed "stuff"

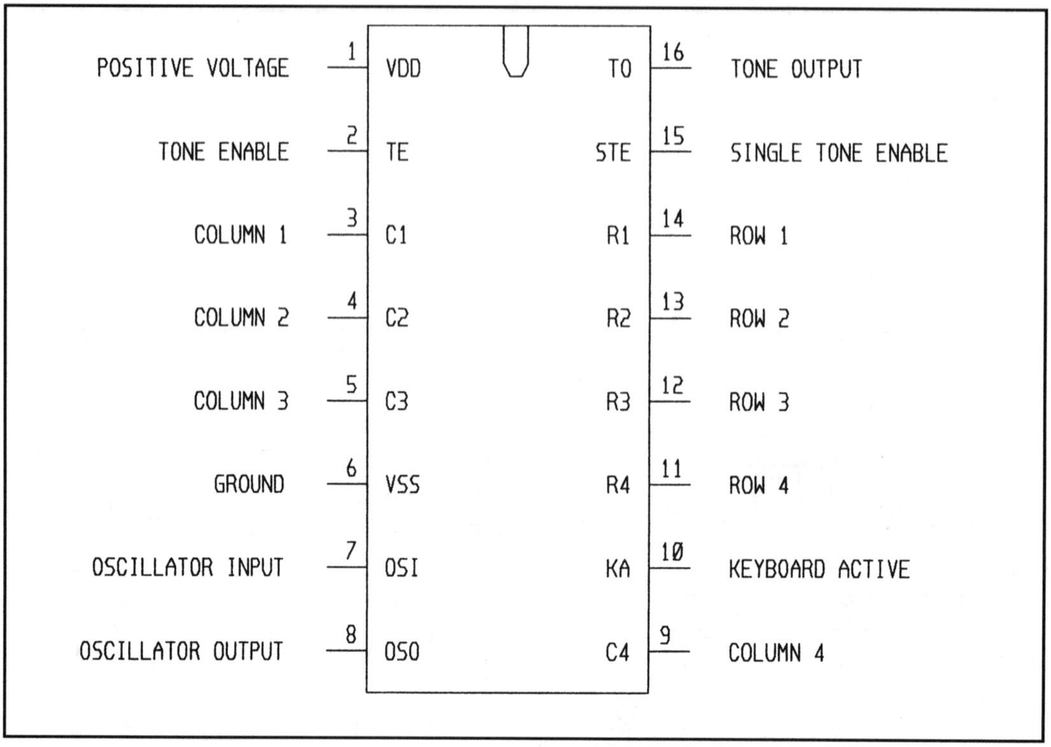


**Figure 6-7.** TCM5089 pinouts.

**Photo 6-7.** The stuffed board of the auxiliary speakerphone.

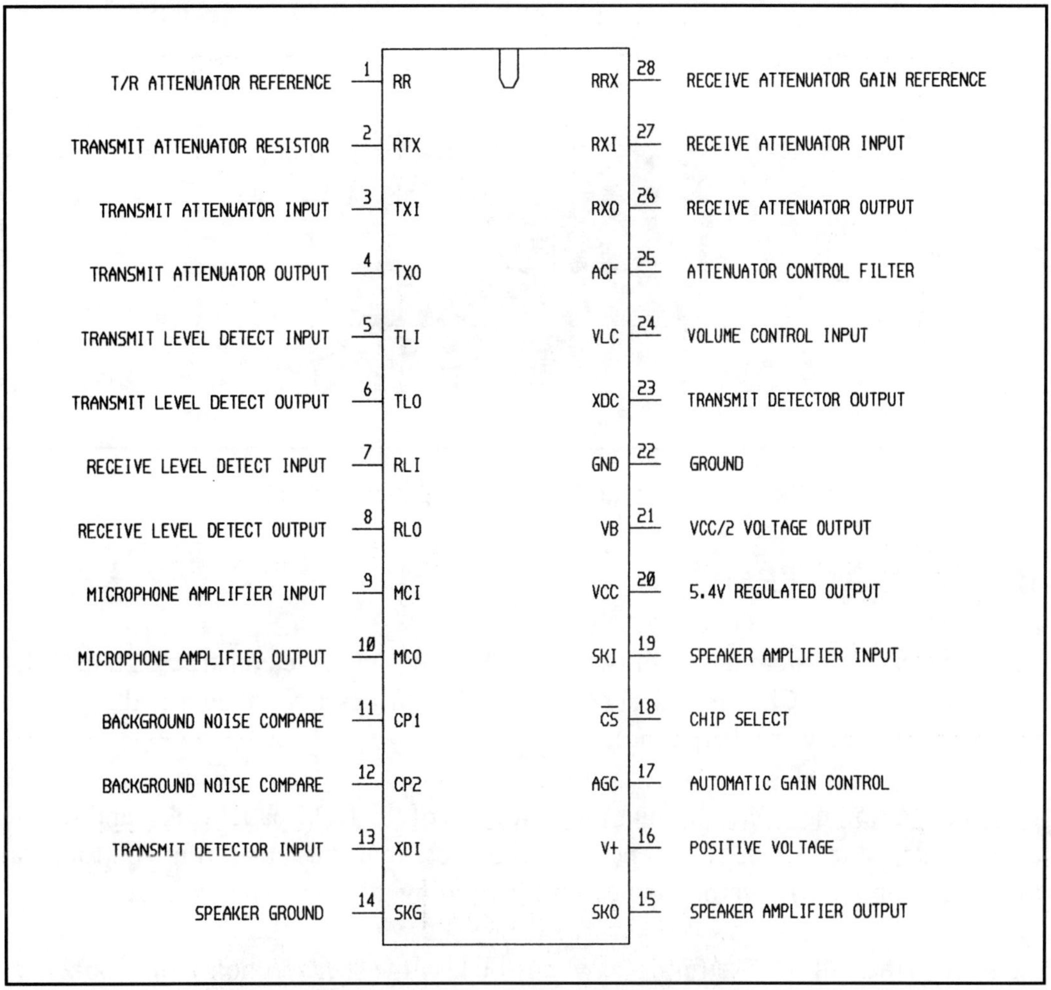


**Figure 6-8.** MC34018 pinouts.

(volume control, switches, mic, etc.), and if desired, label the case with dry transfer lettering.

It is always a good idea to spray a light coat of clear enamel over the lettering, as without protection it is especially susceptible to damage. Once that is dry, the case parts can be installed and the board secured in place.

Since there are no internal adjustments, at this point the case could be closed. However, I recommend running a circuit test before doing that — just in case you need to check anything inside.

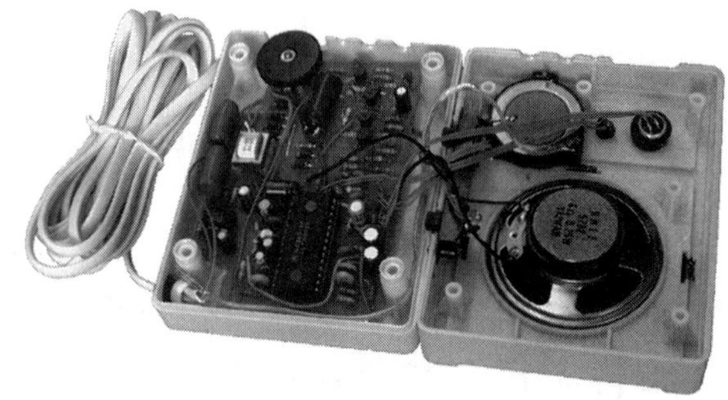

**Photo 6-8.** The auxiliary speakerphone in its case.

## TESTING AND OPERATION

You can test this system in one of two ways. First, if you have constructed the "Telephone Tool Box" in Chapter 8, you now have all the necessary voltages and signals to challenge your new speakerphone.

In this scenario, insert the modular plug into one of the RJ11 panel jacks, and set the appropriate switch to the ring generator (RG) position. Press the "ring" button, and the speakerphone will warble its head off, assuming all is working correctly.

Next, go to the "48 volt" setting, and when you take the speakerphone "off-hook" you will be able to hear the sidetone effect; that is, when you speak your voice will be heard over the speaker. Additionally, the meter reading will indicate a drop from 48 volts to around 5 volts. This demonstrates that the system is placing the proper load on the line.

As for the second testing method, this involves putting the phone on the genuine telephone line. That, of course, is done by plugging it into an active telephone jack. Bear in mind, though, that if something is drastically wrong, you may well damage the speakerphone, or cause problems on the line.

But, never fear! If you have constructed this device in accordance with this chapter, you'll have nothing to worry about. Just be sure to put the fire out as soon as possible.

No, not really! I'm simply trying to add a little humor to what sometimes becomes a very dry subject. Ha, Ha! In actuality, there is very little that can go wrong or cause line interference. However, "Ma Bell" can get downright cranky if you foul up her lines, so it is always something to think about.

Back to business. If the speakerphone is functioning properly, you will hear a dial tone and notice the sidetone effect whenever the device is in the "off-hook" condition. Also, the LED will light. In the "on-hook" position, the phone will be silent until a ringer signal is sent to the line. At that time, the speakerphone will ring.

Now, I know this stuff is complicated, but bear with me. I'll get you through it. When the phone rings, you should place the hook switch in the "off-hook" position which will allow you to communicate with whoever (whatever) is calling you.

See? When I'm on your side, you've got it licked. I mean, stick with me, I'll make you famous. All right, all right. I'm sensing some ugliness again. Hey, I do the best I can.

## CONCLUSION

OK, let's put the humor aside, and wrap this up. Once completed, tested and operational, I suspect you will find numerous applications for both of these devices. If you need a basic phone for the kids' room, Part One will fill the bill. And, the speakerphone speaks for itself. Uh, excuse the pun!

Like the headset phone in Chapter 9, I have found our second endeavor quite handy any time you need to keep your hands free. You know, when you are typing on a computer keyboard as someone is chewin' at your ear over the phone. Don't you just hate that? I do; well, at least sometimes.

So, if you envision that, or other scenarios that might require hands-free telephone operation, the speakerphone is for you. Also, don't forget the family talks with grandma. This device can really make those conversations more rewarding, as everyone is involved at the same time. And, if grandma is long distance, your pocketbook will thank you.

Gee! Ain't electronics wonderful? All these gadgets to make your life easier. What would we all do without them? Have fun!

# CHAPTER 7

# A 300-BAUD MODEM

What telephone projects book worth its salt doesn't at least touch on the subject of modulator-demodulators (MODEMs)? These "gems" have been around for years, but with the recent interest in computers, especially in the home, they have gained revitalized attention and celebrity.

The purpose of a MODEM (MOdulator-DEModulator) is to convert the digital pulses sent from a computer's "serial port" to audio frequency signals the telephone line can handle. As has been stated throughout this text, the phone system was designed for normal speech (roughly 300 to 3,000 hertz), and anything sent over it must conform to that range.

Thus, the "electronic wizards" needed to find a device that would meet that requirement, and the fruit of their labor is the MODEM. In reality, it is simply a combination analog-to-digital converter (ADC) and digital-to-analog converter (DAC) controlled by software. When "sending," the DAC transforms the digital format to analog audio, and when receiving, the ADC performs the reverse operation.

The software meets prescribed standards of the Consultative Committee for International Telephone and Telegraph (CCITT), while the hardware adheres to the "Bell" criterion. In our case, the CCITT is "V.21" and the BELL standard is "103." 103 is described as "300 Baud, Asynchronous, Full-Duplex over 2-Wire Dial Up or Dedicated Telephone Line." This means the MODEM transmits 300 bit per second, one bit at a time, with start and stop bits to define the message. It also allows for two-way communications over standard telephone lines.

Naturally, our MODEM will do all of this. The software is courtesy of International Business Machines (IBM) in the form of their "sample program" found in the IBM "BASIC" manual. It is listed toward the end of this chapter, and is a simple but effective program that runs under the Beginner's All-Purpose Symbolic Instruction Code (BASIC) language. Code operation will be described shortly.

With these two trusty tools, you will be able to send data over any 2-wire "telephone capable" line. Now, this device probably won't get you on the Internet (baud rate is only 300), but it will provide a solid foundation, and hopefully understanding, of modulator-demodulator technology. And, it will come in handy for intercomputer communication or local area networks (LANs).

So, without further delay, let's find out how our MODEM works, and how to build it. The finished product is small and will prove its worth, especially with that old Commodore VIC 20 or C-64 you bought at the flea market.

## THEORY

The circuit revolves around Texas Instruments' TMS99532 MODEM chip, and as stated, this is a blend of specialized serial (one bit at a time) digital-to-analog and analog-to-digital converters. Add to this a clock and carrier detect section, and you have one "handy-dandy" MODEM.

This configuration and software comprises what is known as an "unintelligent," or "dumb" terminal. "Dumb" in the sense that it is manually controlled. But, don't let that discourage you! This device is everything "intelligent" systems are, except for the automatic receive/activation feature.

Block diagram *Figure 7-1* illustrates the TMS99532 layout. As seen, the chip incorporates transmitter and receiver sections, a clock, a carrier detector and digital interface. One disadvantage is the need for a triple-voltage power supply, but that is easily accomplished with zener diodes.

Add an operational amplifier and telephone line interface, and *VOILA!*, a compact full-duplex communications system. All in all, a respectively modest recipe for this dish. Or, should I call it dessert! I'll let you decide.

Referring to *Figure 7-2*, the "line interface" should look familiar. Tip and ring come in through an RJ11 modular jack, voltage spikes are suppressed by metal-oxide varistor MOV1 and capacitor C1 blocks direct current (DC).

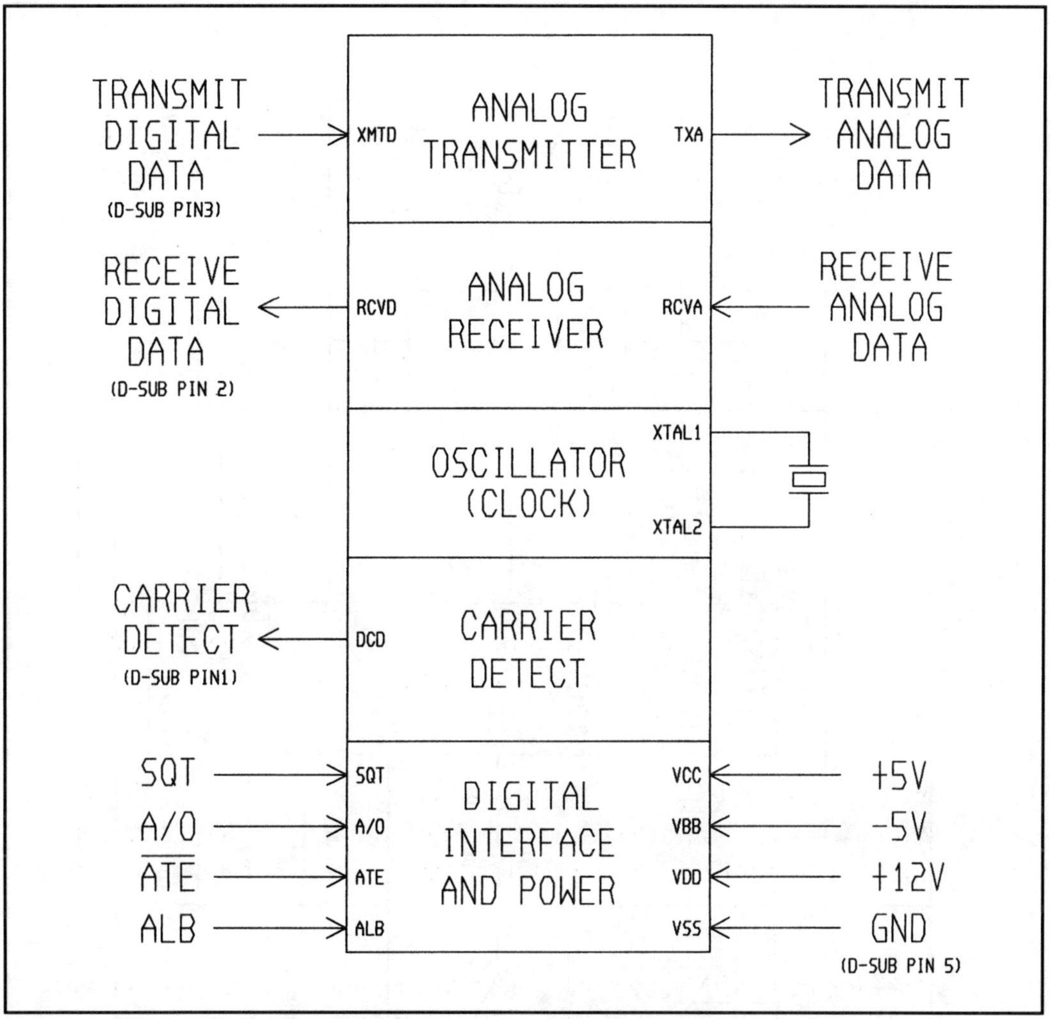


**Figure 7-1.** TCM99532 block diagram.

Matching transformer T1 adjusts the circuit to the 600-ohm telco standard, and zener diodes D1 and D2 clip the 90-volt ringer signals. By matching impedance, T1 insures the system "seizes" the line, and also provides isolation from the telephone "local loop."

Operational amplifiers U1a and U1b (both halves of an LM358 dual-amp chip) are configured as inverting amps to boost the incoming and outgoing analog signals. The various discrete components are part of these basic amplifier circuits. For example, resistors R2 and R6 set gain, and networks R3/C2 and R7/C7 perform input or output coupling.

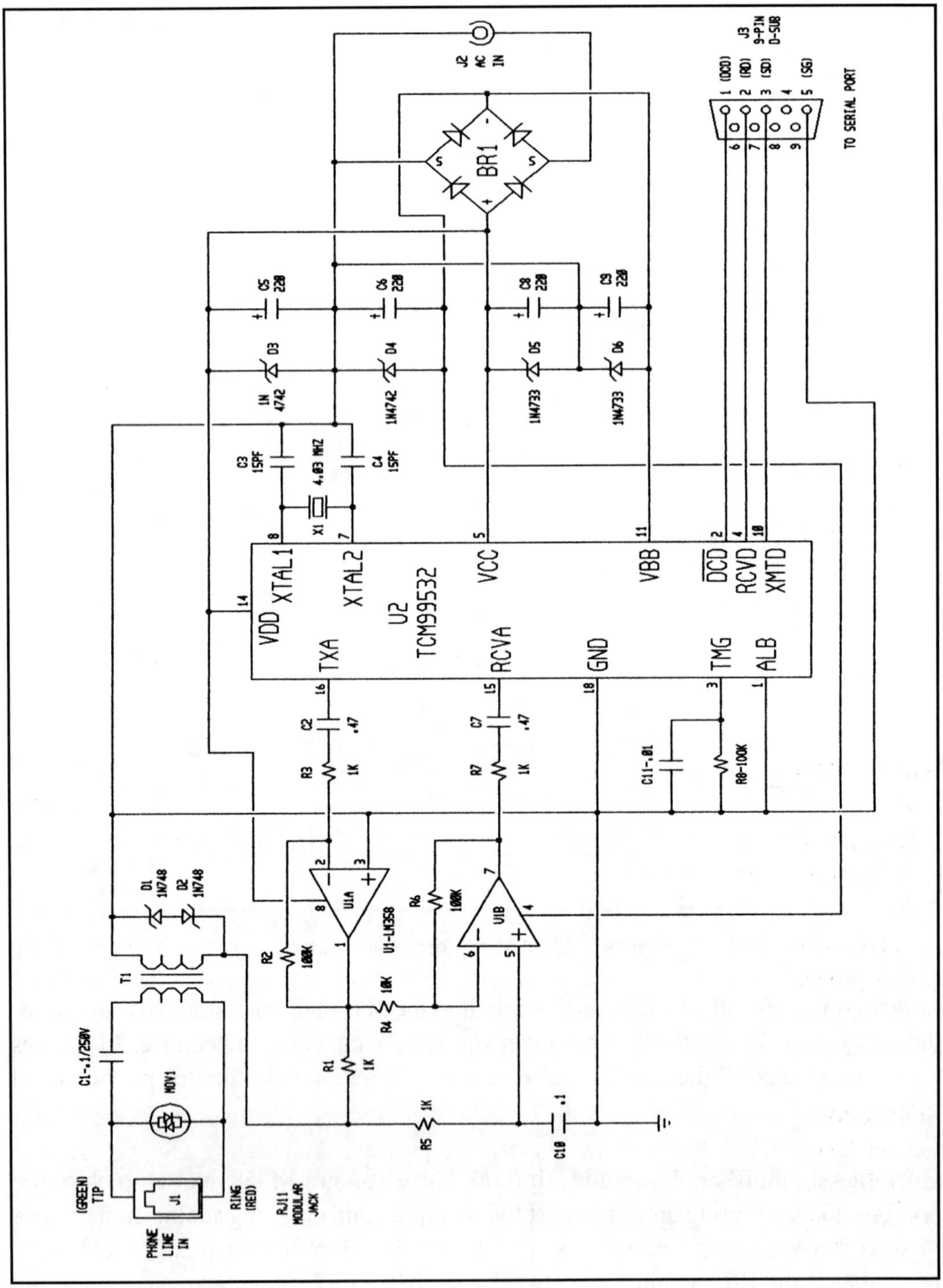


**Figure 7-2.** 300-baud modem.

U1a handles transmission onto the line, and U1b accommodates analog data coming in. Digital information from the computer is managed directly by the TMS99532 (U2).

Speaking of the 99532, an internal clock needs only a crystal and a couple of capacitors to generate the timing. X1, a 4.032-megahertz quartz crystal, is connected to U2 pins 7 and 8, and two small 15-picofarad capacitors are included to "jump start" the internal oscillator.

Since U2 requires three voltages, a zener diode supply is employed to produce the positive 12 volts and positive/negative 5 volts. Op-amp U1 needs a positive/negative 12 volts, so the latter is also furnished.

This regulated power is accomplished in the following fashion. By connecting two zeners in series, the center anode-cathode junction acts as the ground rail, while the unconnected cathode and anode provide the positive and negative voltages, respectively.

A rectified voltage greater than the value of the zener diode is applied to the cathode (+) and anode (-) and one side of the AC line is hooked to the ground rail. The result is a dual-polarity, or "split" power supply.

With the MODEM, we need both 5- and 12-volt supplies, hence two sets of zener diodes are employed. D3 and D4, 1N4742s, handle the 12-volt system, while 1N4733s, D5 and D6, furnish the 5-volt dual supply. In each case, two 220-microfarad electrolytic capacitors are used to filter the incoming DC.

That DC is provided by bridge rectifier BR1, with the AC coming from any 13-volt or higher power transformer. Since the amperage requirement for the circuit isn't that high (maybe 150 milliwatts), zener diodes are more than adequate. They can handle up to one amp.

Connection to the computer is handled by just four serial port lines. These are Data Carrier Detect (DCD), Receive Data (RD), Transmit Data (SD), and Signal Ground (SG). Each respectively goes to the TCM99532's Data Carrier Detect (DCD-pin 2), Receive Digital Data (RCVD-pin 4), Transmit Digital Data (XMTD-pin 10), and Ground (VSS-pin 18).

## 300-BAUD MODEM PARTS LIST

SEMICONDUCTORS:

| | |
|---|---|
| U1 | LM358 Low-Power Dual-Operational Amplifier |
| U2 | TMS99532 300-Baud Modem (TI- Jameco, JDR) |
| D1, 2 | 1N748 3.9V Zener Diodes |
| D3, 4 | 1N4742 12V Zener Diodes |
| D5, 6 | 1N4733 5.1V Zener Diodes |
| BR1 | W02G 200V Bridge Rectifier |
| MOV1 | 130V Metal-Oxide Varistor (V130LA2 or Equiv.) |
| X1 | 4.032 MHz Crystal (JDR, Digi-Key) |

RESISTORS:

| | |
|---|---|
| R1, 3, 5, 7 | 1,000 Ohm 1/4W Resistors |
| R2, 6, 8 | 100,000 Ohm 1/4W Resistors |
| R4 | 10,000 Ohm 1/4W Resistor |

CAPACITORS:

| | |
|---|---|
| C1 | 0.1 μF 250V Mylar/Metal Film Capacitor |
| C2, 7 | 0.47 μF Mylar Capacitors |
| C3, 4 | 15 pF Ceramic Disk Capacitors |
| C5, 6, 8, 9 | 220 μF Electrolytic Capacitors |
| C10 | 0.1 μF Mylar Capacitor |
| C11 | 0.01 μF Mylar Capacitor |

OTHER COMPONENTS:

| | |
|---|---|
| T1 | 600-to-600 Ohm Telephone Matching Transformer |
| J1 | PCB Mount RJ11 4-Conductor Modular Jack |
| J2 | Coaxial Power Jack (to match Power Supply) |
| J3 | Female 9-Pin D-Sub PCB Jack |

MISCELLANEOUS:

PCB, Power Supply (Wall Transformer, min. 13V), IC Sockets, Solder, Hookup Wire, Telephone Cable w/RJ11 Plugs, Serial Cable, Case if desired.

At the other end, each of these lines terminates in the following D-Sub (RS-232) plug pins: DCD to pin 1, RCVD to pin 2, XMTD to pin 3, and VSS to pin 5. For a "dumb" terminal, the remaining five serial lines are not needed. They are only used for automatic, or "smart" terminal operation.

Thus, as seen, most of the work is done by the TMS99532 integrated circuit. With a little help from its friends, it is able to set up a full-duplex, asynchronous communications link.

## CONSTRUCTION

For this project, I DO recommend a printed-circuit board (PCB) more arduously than usual. Due to the fact that we are dealing in audio signals, a PCB helps prevent extraneous noise from entering the system. This is exceedingly important with data transfer, as noise can be mistaken as information, and cause unwanted, and unnecessary, transmission problems.

So, for your convenience, I have include the PCB foil pattern (*Figure 7-3*) and a parts placement diagram (*Figure 7-4*). These will allow you to construct a "trusty," "field-tested" and "combat worthy" board not at all susceptible to noise absorption. Hey, I'm always lookin' out for you!

With the PCB in hand, first install jumpers and IC sockets. Follow this with the resistors, capacitors and diodes. Last come the crystal, bridge, MOV, transformer and jacks. This is a "tight" board, so be especially careful not to create any "oh so sinister" solder bridges and/or cold joints.

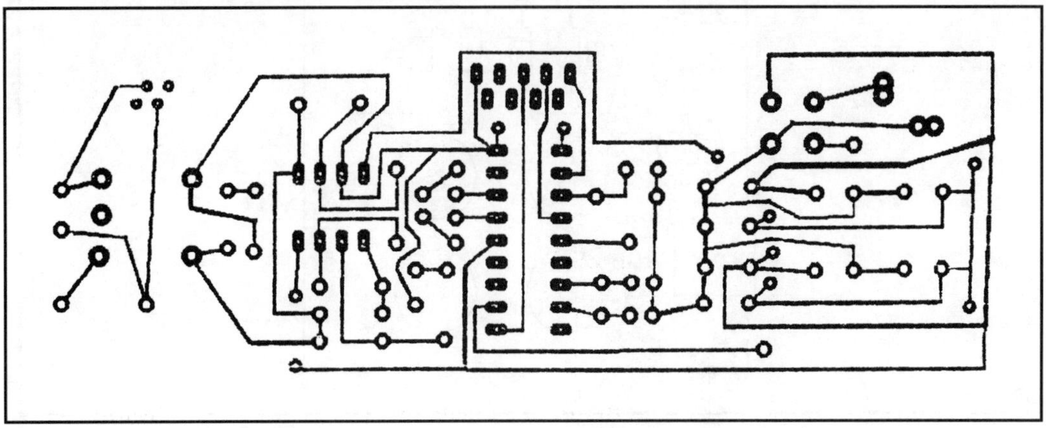

**Figure 7-3.** 300-baud modem PCB foil pattern.

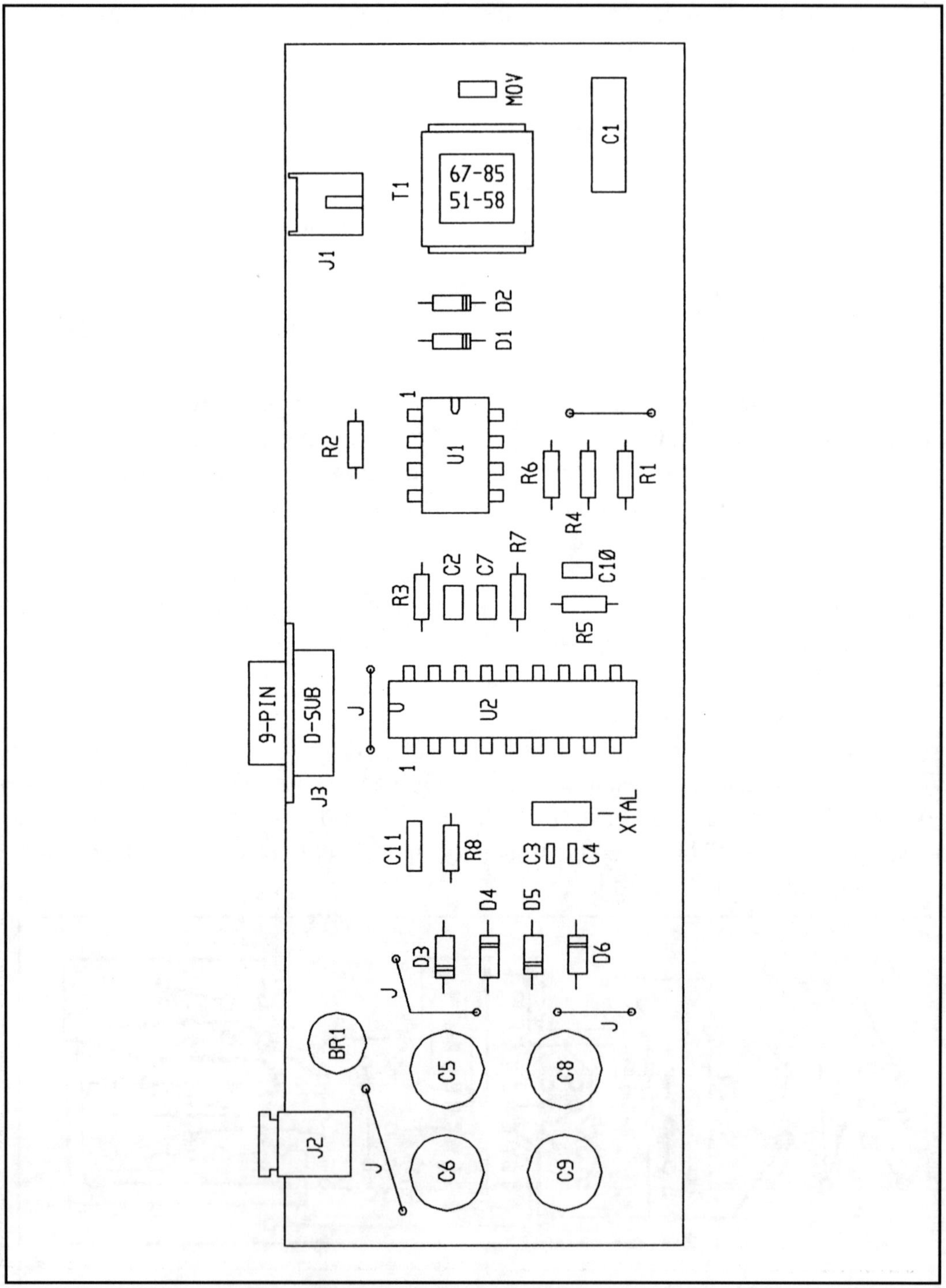


**Figure 7-4.** 300-baud modem parts placement.

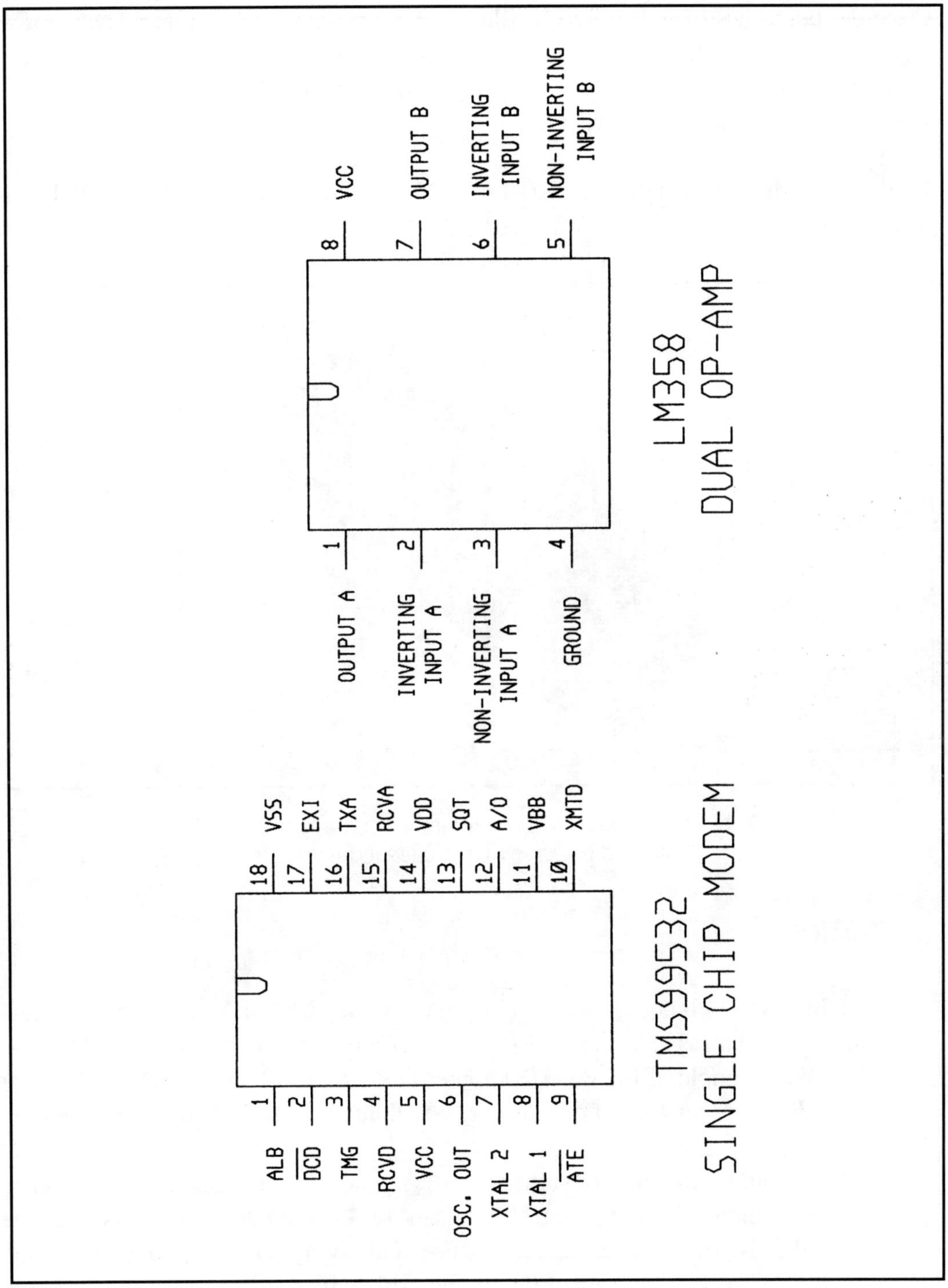


**Figure 7-5.** Modem IC pinouts.

Once the board has been "stuffed" (filled with parts), apply the external AC power and check for the following voltages at the following places: positive 12 at pin 14 of U2 and pin 8 of U1; positive 5 at U2's pin 5, and negative 5 at pin 11; and finally, pin 4 of U1 should show a negative 12.

If all is well, disconnect AC and install the LM358 and TMS99532. The MODEM is now ready to use, so let's hook it up and run a procedural test.

**Photo 7-1.** The completed 300-baud modem ready to use. This board could be housed in a case if desired.

## OPERATION

The first thing you will need is another computer with a MODEM. A faster baud rate is usually acceptable, as most units will "drop back" to whatever transfer speed they're receiving. For example, if the other computer has a 14,400-baud MODEM, when it detects the 300 bps "tempo" of our device, it will automatically adjust to that rate.

OK, with a friend's computer, or someone else's, ready and waiting, attach the serial port of your computer to the 9-pin D-Sub jack on the MODEM. Do this with the computer OFF. Next, power up the MODEM and connect it to the telephone line through the RJ11 jack. A section of telephone cable with modular plugs on each end will do nicely.

Now, "fire up" your computer, and run the IBM software. As stated, the code runs under BASIC, and will have to be typed in the first time you use it. However, saving it to a floppy disk or hard drive makes it ready for action the next time around.

The screen will go blank except for a flashing cursor at the upper left corner — that's provided the software ran correctly. If not, the program will remain on the monitor, and BASIC will take you to the line in question. The problem is usually a typo, so check the code listing against the text you have entered.

Also, with some of the more recent versions of BASIC, like QBASIC, a "pop-up" box will indicate the trouble. This box includes a "help" option that will describe a remedy.

For your next trick, dial the phone number at the other computer, establish connection and start typing information. If all is well, the computer at the other end will receive and display said info. This step, of course, is done after you have obtained a blank screen and cursor.

Incidentally, for the program to work, the MODEM must be connected to the computer. If you try to run it without connection, you will get the "silent screen of death"; you know, a system lockup that requires rebooting to restore computer operation. This is because the system is unable to detect the necessary control signals.

The "reboot" procedure is also known as the "three-finger shuffle" where you press the CTRL, ALT and DELETE keys at the same time. The result is a reset of the system. Or, if your computer has a "reset" button, that will work too.

And, as they say, that is all there is to it. In anticipation of your next question! No, I don't know who "they" are (is?), but there must be someone out there somewhere who says stuff like that. Maybe "they" will write in.

## THE SOFTWARE

By providing the necessary control signals, this software program allows for operation of less complex hardware. The following is furnished to help you understand how the IBM program does its job. While this is elemental "dumb terminal" code, it does the trick for our purpose.

Lines 5 to 60 are all basic "housekeeping" chores, like setting the screen to monochrome with a 40-character width. Additionally, they clear the screen and check to see if all files are closed.

Lines 70, 90 and 110 disable the "soft" keys and define the variables and indicators. 130 specifies the "ON-OFF" status (transmission of data), while 150 sets up communications on the COM1 port at 300 baud, even parity and 7 data bits.

160 opens the COM1 port; however, I added ",CS,DS,CD" to this line. This resolves the problem of program lockup on line 160 due to the absence of the "clear to send" (CS), "data set ready" (DS), and "carrier detect" (CD) signals.

Since we are working with a "dumb" terminal, those signals are not present, but the program looks for them anyway. When it does not find them, it considers that omission an "error" and halts everything. By adding the text and punctuation to the code, it is told not to check for (ignore) them, and all is well.

Lines 170 to 200 set up the screen, and turn on the cursor.

400 handles two tasks. First, it defines the XOFF variable, and second, it sends the program to the 9000 message line if an error is detected.

Lines 500 to 575 enable the keyboard, and keep tabs on the buffer. If the buffer becomes more than half full, a message is sent to the host computer to stop transmitting. 530 keeps checking the buffer. If it is empty, new data can be sent. If not, it moves on to line 570 which continues the "pause."

Line 590 controls character flow by moving as many characters as possible from the buffer to the program. 620 through 645 detect and manage linefeed commands, while 660 sends data to the monitor and checks the buffer for incoming characters. If they are arriving faster than the system can handle, line 570 is again invoked to pause the communications string.

Statement 690 determines when it is time to continue the transmission. When XON is detected, the keyboard is scanned for input.

And last, 9000 handles error messages. When an error is found, this line will send the error number to the screen, thus telling you where the problem is located.

This is a simple program to say the least. But, don't throw bricks at it just yet. For full duplex communications from a so-called "dumb" terminal, this program will furnish the needed control. My sincere thanks to IBM for including this software in their BASIC handbook.

## Program Listing

```
5 REM   table 7-1
10 REM   dumb terminal example
20 'set screen to monochrome text mode
30 '    and set width to 40
40 SCREEN 0, 0: WIDTH 40
50 'turn off key display; clear screen
60 '   make sure all files are closed
70 KEY OFF: CLS : CLOSE
80 'define numeric variables as integer
90 DEFINT A-Z
100 'define true and false
110 FALSE = 0: TRUE = NOT FALSE
120 'define the XON, XOFF characters
130 XOFF$ = CHR$(19): XON$ = CHR$(17)
140 'open communications to file number 1,
150 ' 300 bps, EVEN parity, 7 data bits
160 OPEN "COM1:300,E,7,,CS,DS,CD" FOR RANDOM AS #1
170 'use screen as a file, just for fun
180 OPEN "SCRN:" FOR OUTPUT AS #2
190 'turn cursor on
200 LOCATE , , 1
400 PAUSE = FALSE: ON ERROR GOTO 9000
490 '
500 'send keyboard input to com line
510 B$ = INKEY$: IF B$ <> "" THEN PRINT #1, B$;
520 'if com buffer is empty, check key in
530 IF EOF(1) THEN 510
540 'if com buffer more than 1/2 full, then
550 'set PAUSE flag to say input suspended,
560 'send XOFF to host to stop transmission
```

(Continued Next Page)

```
570 IF LOC(1) > 128 THEN PAUSE = TRUE
575 PRINT #1, XOFF$
580 'read contents of com buffer
590 A$ = INPUT$(LOC(1), #1)
600 'remove linefeeds to avoid double spaces
610 'when input displayed on screen
620 LFP = 0
625 'look for linefeed
630 LFP = INSTR(LFP + 1, A$, CHR$(10))
640 IF LFP > 0 THEN MID$(A$, LFP, 1) = " "
645 GOTO 630
650 'display com input, and check for more
660 PRINT #2, A$; : IF LOC(1) > 0 THEN 570
670 'if transmission suspended by XOFF,
680 'resume by sending XON
690 IF PAUSE THEN PAUSE = FALSE
695 PRINT #1, XON$;
700 'check for keyboard input again
710 GOTO 510
8999 'if error, print its number and retry
9000 PRINT "ERROR NO."; ERR: RESUME
```

## CONCLUSION

Understanding MODEMs and their operation is, to say the least, important in this day and age of computers. Those digital beasts play a significant role in our day-to-day lives, and for us electronics enthusiasts, in our hobby and/or work.

While not an especially complicated device, the MODEM is interesting. The ability to change digital data into analog, then back again, and both transmit and receive information, makes for fascinating study — and, for a pragmatic addition to any computer. Furthermore, as we have all seen in the recent years, MODEMs can open up an entirely new source of information to anyone with "home computing" capabilities.

Hopefully, this chapter has given you an insight into this sometimes cryptic world. Here, as is so often the case these days, a marriage between "hardware" and "software" (with a little technical ingenuity thrown in for good measure) renders a marvelous solution to a problem once thought unsolvable. However, few dilemmas escape the imagination and expertise of the electronic geniuses, and asynchronous communications was no exception.

Well, there you have it. Try this one out, as it will establish a firm foundation for future experimentation with MODEMs. Right now, the "pickin's are slim" regarding readily available MODEM chips, but that will change.

And as it does, you will be ready for all those fancy ICs that spill out into the retail semiconductor market. After all, one MODEM is basically the same as the next, functionally speaking. There may be differences in speed or features, but with the knowledge gained here, working with those newcomers will be a day at the beach!

Enjoy!

# CHAPTER 8

# THE TELEPHONE TOOL BOX

Here is a valuable companion to Chapter 2's Telephone Lab. Whereas the lab helps you with your prototypes, the "tool box" helps you test those projects. Additionally, it lets you examine the condition of virtually any telephone.

Inside the box you will find all the necessary electronics to generate a dial tone, a ringer signal, a positive 12 volts and an adjustable power supply for the standard 48-volt DC tip-to-ring potential. With this in hand, your projects and/or phones can be tested without actually placing them on the telephone line.

Each section is relatively simple in design, but serves an important function either in diagnostics, or as a method to confirm proper equipment operation. In this manner, the "telephone tool box" becomes a worthwhile addition to your telephone-related test/work bench.

So, let's take a look at the tool box section by section. As we go along, it will become evident how to use modest methods to achieve effective goals. This piece of gear will also help take some of the mystery out of telephone system operations.

## THEORY

Let's start with the dial tone generator (see *Figure 8-1*) — this should be an old friend to most of us. Here, an LM555 Timer/Oscillator is used in the astable configuration to produce a 400-hertz signal. When that is placed across the telephone's tip and ring lines, the tone is heard in the receiver.

This circuit may be more accurately described as a "dial tone simulator." As you may remember from Chapter 1, a genuine telco dial tone is a tone pair of 350 hertz and 440 hertz. By using a "middle-of-the-road" frequency of 400 hertz, the sound produced will approximate a dial tone, but actually will not possess both individual tones.

Thus, this section can't be used for tests that require an actual dial tone pair. However, this need is relatively infrequent, and when necessary, that drawback can easily be overcome by hooking the device to the telephone line.

As for operation, the LM555 functions as an oscillator controlled by the R/C tuning circuit of C1 and R1/R2. By adjusting R1, a 400-hertz signal is generated, and when that is applied to the telephone input, the result is a tone that sounds like an ordinary dial tone. At least it does to me.

Transformer T1 is a standard audio unit with an 8-ohm primary and a 1200 (CT) secondary. Using the center tap (CT), you have 600 ohms which matches the normal telephone line impedance.

Tuning the 555 oscillator is best done with a frequency counter. However, be sure to regulate the input signal (use a 1N4733 zener diode), as many counters are 5-volt systems and can't tolerate the 12-volt output of this section.

If you don't have a frequency counter, go buy a copy of my *Build Your Own Test Equipment* book. You'll find it in Chapter 11. Just kidding (well, not really...uh)! Actually, the easiest way to set the oscillator is to listen to the tone over the phone itself. When it sounds right, it's tuned.

Again looking at *Figure 8-1*, move east to the 48V adjustable supply. This, of course, generates the nominal tip-to-ring "battery voltage" when the phone is "on-hook."

What we have here is a classic voltage doubler circuit, a classic voltage divider and a classic analog meter arrangement. Sorry, no classic Coke! This circuit is really self-explanatory, but let me run through the fundamentals anyway.

Alternating current from the 24-volt power transformer is introduced at the ground rail and through capacitor C3. Diodes D1 and D2 multiply the voltage by two and rectify it to DC. Capacitor C2 then filters the output for a clean voltage twice that of the input.

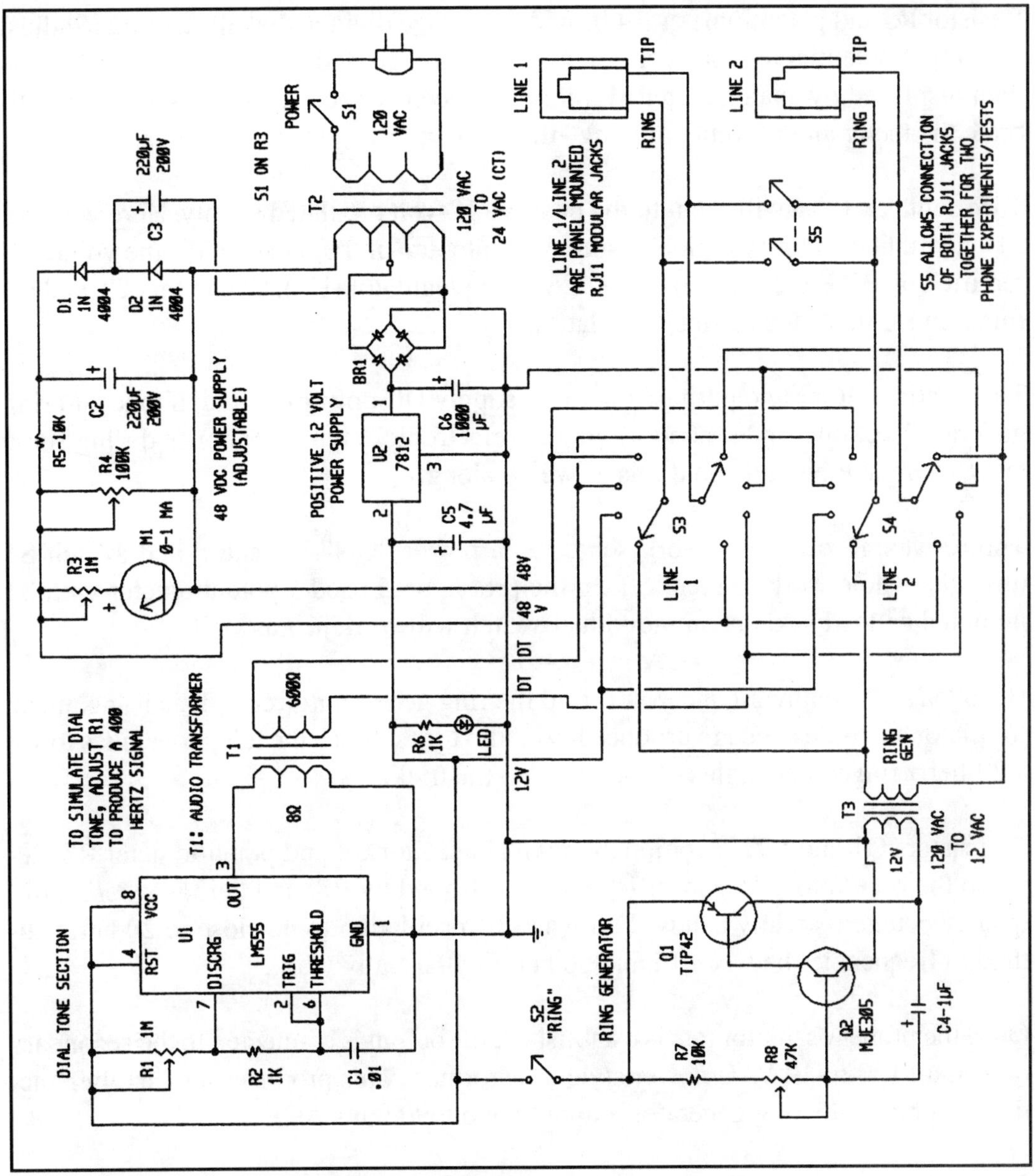


**Figure 8-1.** Telephone tool box schematic.

Caps C1 and C2 should be fairly large (100 microfarads or higher), and at a minimum voltage of twice the input AC. I used 200-volt units, which is overkill, but I had them handy. So, you know, "Don't bite the gift horse in the mouth." Is that right? No, that's not right. Oh well, it's something like that.

Resistor R5 and potentiometer R4 provide a voltage divider that allows you to adjust the output potential to whatever you need (usually around 48 volts). That voltage is then displayed by analog meter M1. Potentiometer R3 acts as a series shunt to calibrate the meter and prevent "pinning" the pointer.

Turn south and you will run into the positive 12-volt regulated supply. Here we have a standard linear configuration. Using the center tap of transformer T2, the voltage is rectified by BR1, then sent to a 7812T voltage regulator (U2). Capacitors C5 and C6 function as filters and prevent oscillation.

This section serves two purposes: first, to supply DC power to the dial tone and ring generator sections; and second, to power "electronic" style telephones during some tests. More will be said about this as we go along.

The resistor R6/LED1 network serves as a power "ON" indicator, and switch S1 turns the whole works on and off. In the prototype, I used a potentiometer with S1 mounted on the back, but an individual switch will work just as well.

Head way out southwest and you'll find the ring generator circuit. This is, again, an oscillator and is designed to produce low audio tones. Since the telephone likes to see a 20-hertz ringer signal, this ol' boy will do the trick.

Transistors Q1 and Q2, in conjunction with capacitor C4 and potentiometer R8, develop the oscillation. The specific rate is controlled by R8, and here is where a frequency counter is really handy. You want the oscillation to be close to 20 hertz, although frequencies between 15 and 50 hertz will usually work.

Once the proper oscillator rate is established, the output is connected to the secondary of a small 120-to-12 VAC power style transformer. This produces an "inverter" action that raises the ring generator output to about 90 volts AC.

That output, in turn, activates the ringer in the telephone. In case you are wondering why power (TO-220 case) transistors are used, the following should explain. I originally used smaller (TO-92 case) units, a 2N2907 and 2N2222, but they became quite hot, especially under load. Thus, I switched to the larger transistors, a TIP42 and MJE305, and that alleviated the overheating problem.

The last section, down southeast (sometimes called "Dixie"), we will designate the "control system." That's a fancy name for three switches and two RJ11 modular telephone jacks, but let's not sell this section short. It allows us to send any one of the four individual outputs to either or both of the RJ11 jacks. And, all that becomes very important.

Through our control system, we can do such things as test single telephones, other telephone equipment and/or hook two devices together. This last scenario permits you to test an answering machine, or establish communications between two phones.

Each rotary switch has its "polls" wired to the tip and ring terminals of one of the modular jacks, and the "throws" connected to the outputs of the four previously described devices. Switch S5 merely wires in parallel the tip and ring of each jack.

So, in this fashion, either or both RJ11's can be supplied with various voltages, a ring signal, dial tone or just talk to each other. The possibilities are staggering. Well, kind of.

Now that you have a firm understanding of how this "thing" works, let's move on to putting it together. I housed the printed-circuit board, power transformer and panel components in a standard project case, but it could also be incorporated in other equipment.

## TELEPHONE TOOL BOX PARTS LIST

| SEMICONDUCTORS: | |
|---|---|
| U1 | LM555 Timer/Oscillator |
| U2 | 7812T Positive 12V Regulator |
| Q1 | TIP42 PNP Power Transistor |
| Q2 | MJE305 NPN Power Transistor (Radio Shack) |
| BR1 | W02G 200V Bridge Rectifier (or equivalent) |
| D1, 2 | 1N4004 Rectifier Diodes |
| LED1 | Red T-1 3/4 Jumbo Light-Emitting Diode |

(Continued Next Page)

RESISTORS:

| | |
|---|---|
| R1, 3 | 1,000,000 Ohm PCB Potentiometers |
| R2, 6 | 1,000 Ohm 1/4W Resistors |
| R4 | 100,000 Ohm Panel Potentiometer |
| R5, 7 | 10,000 Ohm 1/4W Resistors |
| R8 | 47,000 Ohm PCB Potentiometer |

CAPACITORS:

| | |
|---|---|
| C1 | 0.01 μF Mylar Capacitor |
| C2, 3 | 220 μF 200V Electrolytic Capacitors |
| C4 | 1 μF Electrolytic Capacitor |
| C5 | 4.7 μF Electrolytic Capacitor |
| C6 | 1,000 μF Electrolytic Capacitor |

OTHER COMPONENTS:

| | |
|---|---|
| T1 | 8-to-1200 Ohm CT Audio Transformer |
| T2 | 120-to-24 VAC CT Power Transformer |
| T | 120-to-12 VAC Power Transformer |
| S1 | SPST Switch (on R4 in Prototype) |
| S2 | SPST Pushbutton Switch |
| S3, 4 | 2P6T Rotary Switches |
| S5 | DTST Slide Switch |
| L1, 2 | 4-Lead RJ11 Modular Telephone Jacks |
| M1 | 0-1 mA Analog Meter Movement |

MISCELLANEOUS:

PCB Materials, Solder, Paint, Epoxy, Hookup Wire, 120 VAC Line Cord and Plug, Knobs, Dry Transfer Lettering, Hardware, Case, 8-Pin IC Socket, LED Holder (T-1 3/4), Etc.

## CONSTRUCTION

As always (well, almost always), I'm providing a PCB pattern (*Figure 8-2*) for your convenience. Point-to-point wiring will work, but a printed circuit is usually the preferred route.

With the completed PCB in hand, start by installing the jumpers and small components (resistors, capacitors, diodes, bridge, etc.), then the larger caps, transformers and transistors (see *Figure 8-3*). Also, the regulator needs to be heatsinked and mounted to the PCB.

As stated, I put everything in a 7 3/4 inch (196 mm) by 5 1/4 inch (110 mm) by 2 3/8 inch (60 mm) plastic project box. Using *Figure 8-5* as a template, the top, or panel, is prepared by drilling and cutting appropriate holes for the meter, switches, modular jacks, potentiometer and LED. The bottom section needs only a hole for the power cord. With that done, it's time for the paint job.

I painted the bottom section a medium blue and used a light yellow for the top panel. Once the paint is cured, dry transfer lettering is used for labeling the panel. This is followed by a LIGHT coat of clear enamel to protect the lettering.

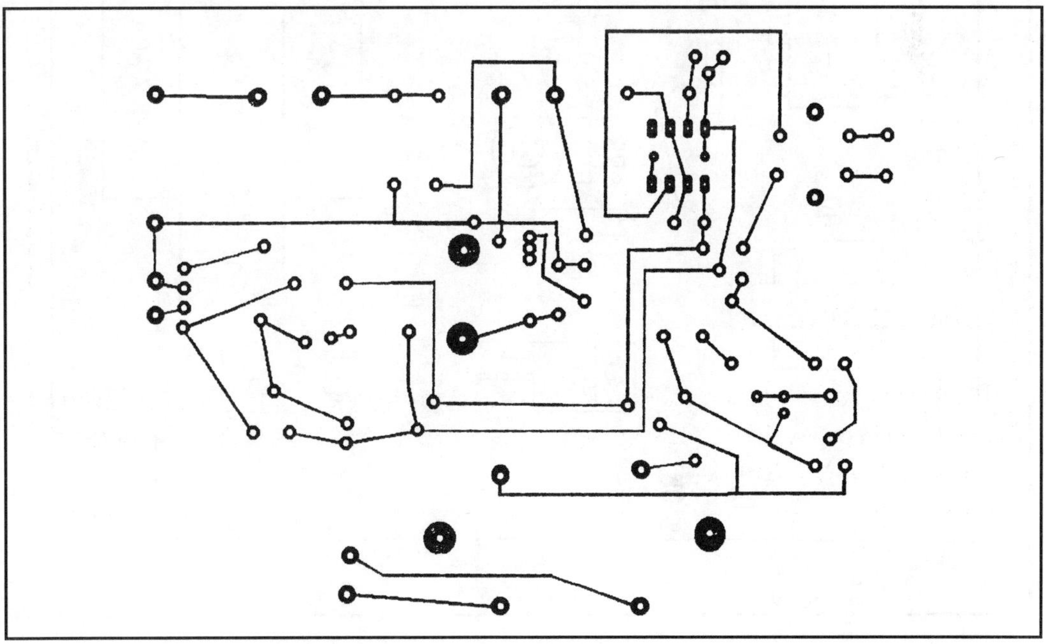

**Figure 8-2.** Telephone tool box PCB foil pattern.

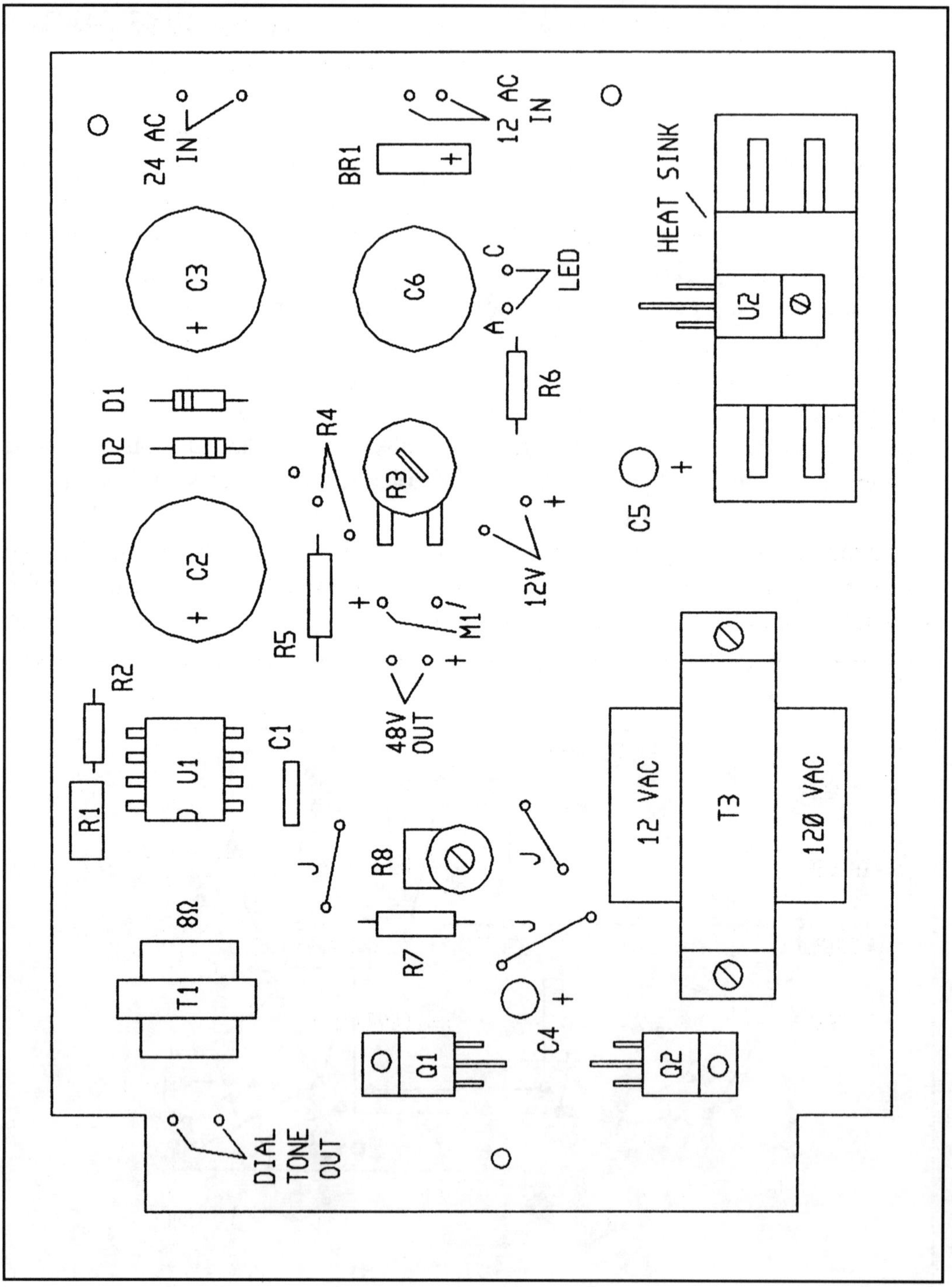


**Figure 8-3.** Parts placement.

I know I have covered this before, but I'm gonna hit it again. Be sure the protective coat is light, as some dry transfer materials will melt from the solvents in the paint. A real mess!

Also, DO NOT use clear lacquer over the yellow enamel. Lacquer will cause the enamel to bubble and crack. I don't mean to belabor these points. I'm just trying to spare you from some of the problems I have encountered in the past. And, save you from the mess they can create.

OK, with the panel ready, install the various components. I used epoxy to hold the RJ11 jacks in place, an LED holder for...you guessed it...the LED, and the rest come with their own mounting hardware. Interconnections between the rotary switches, S5 and the modular jacks can be made at this time. When completed, set the top aside for the moment.

Now is the time to connect appropriate lengths of hookup wire to the PCB. These go to the panel parts, and when I say appropriate lengths, I'm talking about enough wire to reach the panel and leave some working room.

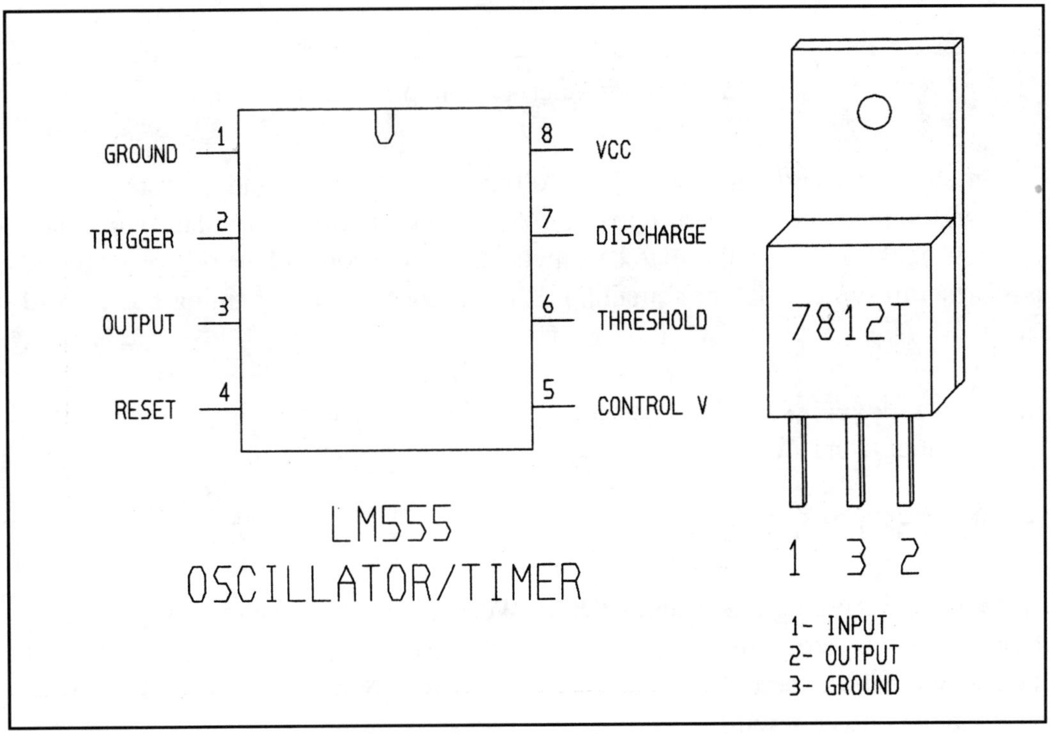


**Figure 8-4.** Tool box IC pinouts.

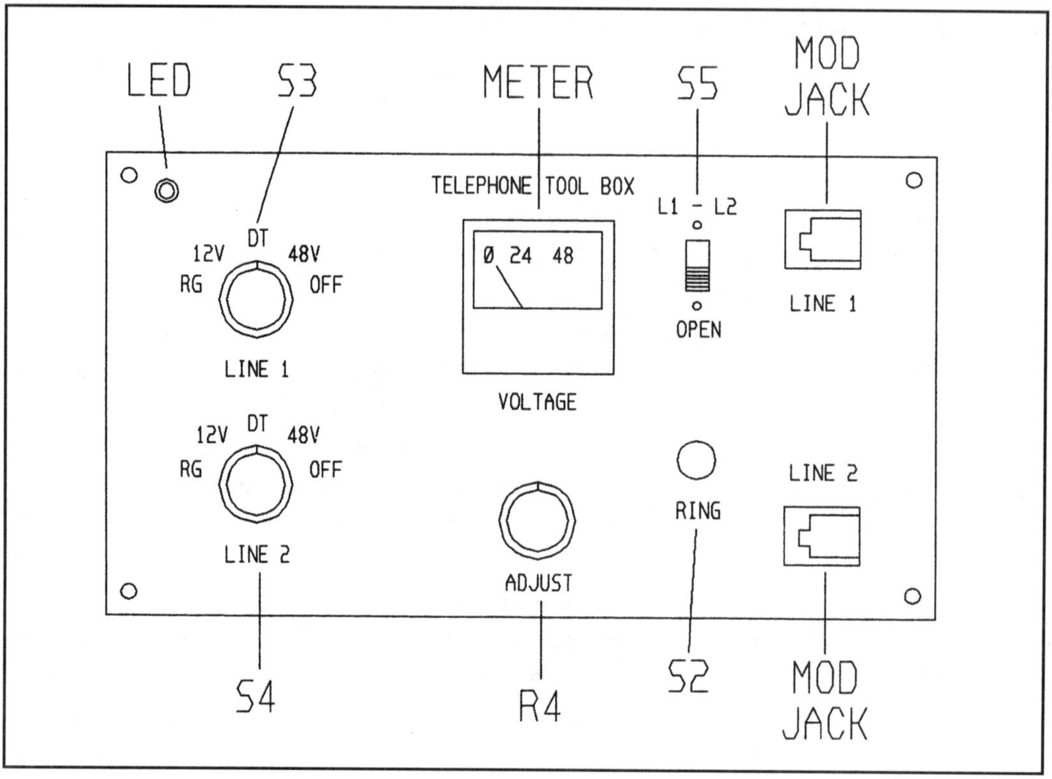


**Figure 8-5.** Telephone tool box front-panel template.

Next, mount the board and transformer to the case with bolts and spacers (for the PCB), and connect the power cord to one side of the transformer. Find the panel, and solder the hookup leads to the switches, meter, etc., then connect the other side of the transformer to switch S1. The remaining S1 lead goes to the remaining power cord lead.

With all that done, it's time to test the system. Don't close the case just yet, as the PCB potentiometers still need to be set.

## TESTING THE TOOL BOX

The first step is to plug the thing in (Duh). Well, maybe not so duh. Many a panic-stricken hobbyist, struggling with an apparently dead circuit, has found great relief when he or she discovered the line cord was unplugged. Now, don't laugh. If it hasn't happened to you yet, it will!

**Photo 8-1.** The stuffed board for the telephone tool box.

Anyway, with the tool box plugged in, activate switch S1. You should see an illuminated LED, and a reading on the analog meter. Next, check voltages at pins 4 and 8 of the LM555, the output (pin 2) of regulator U2 and the collector of transistor Q1. All should read between 11.95 and 12.05 volts DC.

For the 48 VDC generator, connect a digital voltmeter (DVM) across the output at analog meter M1. Since most transformers provide higher voltages than their rating, this circuit may well produce voltage in the 50 to 60 volt range. This is where potentiometer R4 comes into play.

Adjust R4 until the DVM reads 48 volts. Now, adjust potentiometer R3 to get the pointer of the analog meter to swing about 2/3 full scale. Mark the faceplate for that position. Next, set R4 for a lower reading, say 24 volts, and again mark the faceplate. Try a 12-volt setting, and maybe a reading above 48 volts, and mark each. These marks allow you to create a new meter faceplate representative of the supply's output.

If all is working correctly, scare up a short piece of telephone cable with a modular plug on one end. Strip the ring (red) and tip (green) leads, and plug the cable into one of the panel modular jacks. Now, as you rotate the respective switch (S3 or S4) the various voltages will appear at the tip and ring.

Now, what follows is not to spook you, create panic or otherwise deter you from working with this project. It is merely to keep you alive. No, just kidding again. All I wanted to say is that you will be dealing with some higher voltages here, so don't be grabbing loose modular cable wires. The potentials aren't deadly, but will nip at you.

Also, check to see if both rings and both tips are connected together when S5 is closed. If everything is as it should be, the testing is completed, done, finished, or as the Hawaiians would say, "au pow, brudder."

## OPERATION

Let's say you just bought a telephone at a garage sale, and you want to find out if it works correctly. Well, you could plug it into your home line and see if it does its thing. Of course, this could wipe out your telephone service.

So, rather than take that chance, try plugging it into the telephone tool box you just built. Sounds smart to me! Actually, in addition to being smart, it's more practical. Using the tool box allows you to test various functions of the garage-sale phone.

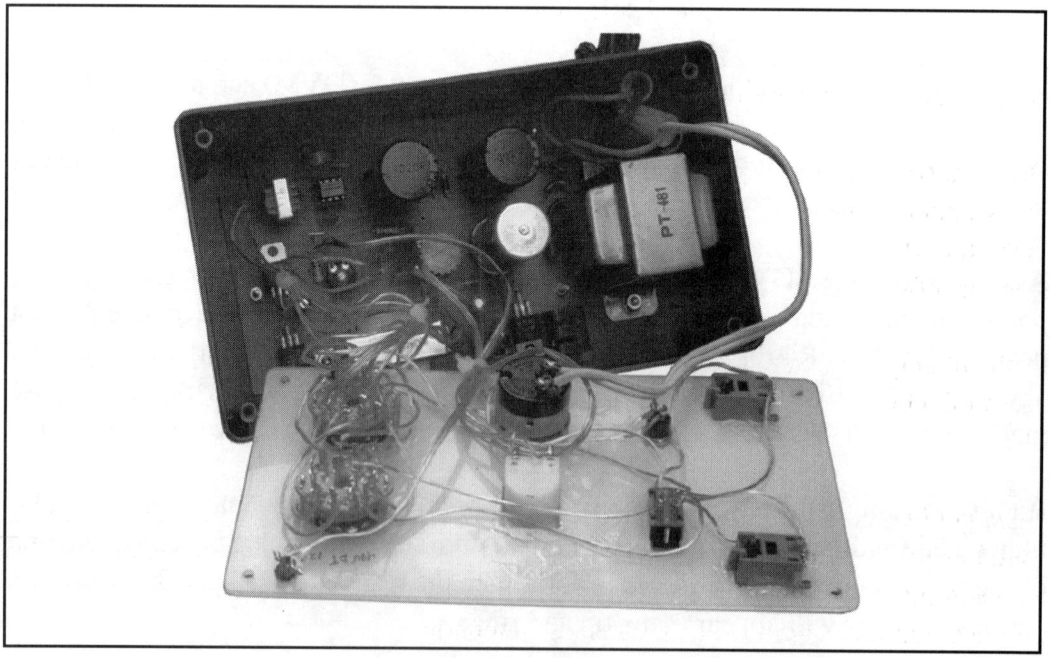


**Photo 8-2.** An interior view of the completed telephone tool box.

With your newfound $3 bargain plugged into either modular jack, set the appropriate rotary switch to dial tone (DT) and turn the tool box on. If the phone is "off-hook" and its receiver section is operating correctly, you will hear the tone. If no tone, then several different gremlins may be rambling around inside that garage-sale gem.

The line cable might be bad, the speaker could be defective or the electronic circuitry could be shot. Any or all of these factors are possible reasons for the telephone failure. Or, it could be a single component such as a leaky transistor or open diode. The definitive answer will require additional exploration.

If the dial tone is present, change the switch to ring generator (RG) and put the phone "on-hook." Press switch S2, and if the ringer circuit is good, the telephone will ring. If it doesn't, check out the components in that section.

For electronic phones, set the rotary to 12 VDC. With a working telephone you will be able to hear your voice in the speaker when you talk into the microphone. This is known as "sidetone" and is a passive operation that lets you know your voice is being transmitted over the line.

Also, with some electronic telephones, 12 volts will allow you to activate the DTMF tone pairs. These can be heard over the speaker, and this indicates the "touch tone" pad is functioning. However, many electronic phones, especially the cheapies, will not activate the tone dialing with the 12-volt source.

Older pulse style telephones will require the 48-volt power supply, so if your garage-sale wonder is one of those, set the rotary switch to the 48-volt position. Be sure to set the voltage to 48 with the phone on-hook. When off-hook, you will be able to hear the sidetone effect. Also, the DTMF tones will be heard clearly at this voltage.

With either type device, you will notice the traditional voltage drop when the phone is taken off-hook. For proper operation, the meter should fall to about 5 volts, or near zero.

Now, get two electronic phones, and connect one to each RJ11 jack. Set both rotary switches to 12 volts, close S5, and you should be able to send dialing tones/pulses between the two phones. By manipulating the switches you can ring either phone, produce dial tones and so forth. On some telephones, you may even be able to carry on a conversation.

To test other telephone equipment, all the typical rules apply. Let's take an answering machine (AM) for example. Plug the machine into one of the jacks, set the rotary to ring and determine the number of rings the AM is set for (2, 4, 6 rings, etc.). Now, press S2 that many times, and the answering machine will seize the line and play its message. When sending the ring signals, hold the switch down anywhere between 1/2 and 2 seconds for each ring. This will simulate the normal telco operation.

Hook up a phone to the other RJ11 and set that jack for 48 volts, close switch S5 and when the answering machine "beeps" you will be able to leave a message. Also, if you have a telephone on or plugged into the machine, you will be able to test the "pickup" function. If all is well, the telephone will seize the line, taking it away from the answering machine.

## CONCLUSION

And, so it goes. With the tool box on your bench, you will be able to analyze the operation of most telephone equipment. Of course, this device also lets you evaluate the success of your own prototypes. Trust me, projects involving the telephone system can be challenging and temperamental. But, having this unit available will make those endeavors far less frustrating.

As stated at the outset, the tool box affords the necessary potentials and signals to operate telephone equipment without physical connection to the phone line. In that lies one of the most beneficial aspects of this device. Trying to test phone equipment and/or prototypes using the telco system can be disheartening, as well as annoying to the phone company.

As a matter of fact, so annoying to the phone company that they can get downright testy about the whole thing, even to the point of threatening to discontinue your service. So, to stay on the good side of your local telco, and to make life easier on yourself, give this gadget a whirl and a twirl. Or a twirl and a whirl. Well, you know—try it out.

Matched with the telephone lab, you will be designing fantastic phone equipment in no time at all. OK, if nothing else, you can always turn it into a "field phone" for your kids, or your cousin's kids, or your cousins or somebody. Have fun!

## CHAPTER 9

# THE "HANDS-FREE" HEADSET TELEPHONE

If you spend as much time in front of a keyboard and computer as I do, you will love this project. What we are going to build is an electronic telephone with a headset instead of the conventional handheld receiver.

The headset contains both the microphone and the speaker (earphone), and in this fashion, you will be able to keep both hands free while conducting a phone conversation. You know, free to type on the keyboard.

Actually, free to do whatever you were doing when the phone rang. Or, free to do something that comes to mind after you have started the conversation. Very handy, indeed!

Now, at first glance, this might sound like more of a frill than a necessity. But, once you have constructed this device and put it to use, I believe its merit will become striking. For example, sized slightly larger than a pack of cigarettes, it's extremely portable — an attribute that adds to its versatility.

Thus, without further hype, let's dive into this project head first. The sooner you build this telephone, the sooner you can start using it (I'm a master at stating the obvious.), and I think you might learn a little more about telephones along the way.

## THEORY

As you have probably already gathered, most telephones are basically the same, at least in function. And this device, similar to the desk phone in Chapter 6, doesn't breach that observation. The same principal components—a ringer, dialing system and audio circuitry—are all present. The difference is that each is accomplished electronically.

This approach allows the phone to be both compact in size and reliable, as there are few mechanical parts to wear out. However, the electronic strategy does require slightly different methods to accomplish the task. This is especially true in terms of the "sidetone" and modulation functions.

Looking at *Figure 9-1*, you will see that the heart and soul of the phone is one integrated circuit (IC) and seven transistors. Each, of course, has a special role in the "BIG PICTURE," so let's take a look at each individual section and how it performs.

To start with, the telephone line is introduced at the "tip" (green wire) and "ring" (red wire) connections. Each is directed to one of three segments of the circuit. Let's take the "ringer" section first.

It is made up of transistor Q1, capacitor C1, a three-lead piezo element, switch S2 and resistors R1, R2 and R3. As you may remember, the ringer is a 20-hertz AC signal, usually about 90 volts in intensity, superimposed on the normal 48-volt battery voltage.

To detect this signal, we configure transistor Q1 as an oscillator triggered by the positive half of the AC cycle. The piezo element is wired as part of this circuit to audibly announce the presence of the ringer signal (Hello! Answer the phone!).

Capacitor C1 is used to block DC, thus keeping the ringer from responding to the battery voltage. Switch S2 merely disconnects one side of the line, in this case the "ring," which effectively shuts off the ringer. That's when you want peace and quiet, or something like that.

OK! So much for the ringer. The next connection to the line is the "hook switch." For our project, this is a DPST slide switch, and as *Figure 9-1* reveals, connects the telephone line to rectifier diodes D1 through D4.

Since there are times when the battery voltage will change polarity, the diodes are used to maintain fixed positive and negative voltage terminals. This protects the IC and transistors, as well as any other polarity-sensitive components, against damage due to line changes. Incidentally, a bridge rectifier can be used instead of four diodes, but must be rated at 400 volts.

Once the voltage has passed through the rectifiers, another diode is engaged to regulate said voltage. Considering the 48-volt battery rating, as well as the 90-volt ringer signal, and considering the IC will burn up at potentials much above 5 volts, regulation is definitely in order. This is accomplished by connecting a zener diode across (in parallel with) the rectifier outputs. A 1N4733 restricts the incoming voltages to about 5.1 volts, providing an ideal working voltage for the circuit.

Next in line is the light-emitting diode LED1 and resistor R4 network. These are employed to indicate the phone is off-hook and functioning. Nothing phenomenal about this part.

And, that leads us to the dialing section of the system. That, of course, is integrated circuit U1. This gem of the electronic world converts strokes on a numerical keypad into Dual-Tone Multifrequency (DTMF) tone pairs. And those, when routed to the central office, are interpreted as individual digits.

There are a number of tone generator (encoder) ICs available, but I went with a Holtex HT9205A mainly because I had one on hand. They all accomplish the same goal—that is, generating tone pairs—so any of them would work in this application.

As for some specifics on the HT9205 (see *Figure 9-4*), it is crystal-controlled (pins 6 and 7) using a standard 3.579545-megahertz color burst crystal. This provides device timing and maintains the frequency precision required by your telco.

Pins 11 (HKS) and 14 (MODE) are both tied to ground for the following reasons. Taking pin 11 low activates the keypad, and grounding "mode" provides DTMF output. If pin 14 is connected to the positive voltage, the chip outputs "pulses."

Hence, you could add a third switch that allows either tone or pulse dialing, but the pulse output (PO), pin 12, then has to be connected to a couple of transistors and some resistors. And, the whole thing just complicates the circuit. Since very few areas require pulse dialing anymore, I left that out.

The DTMF output (pin 13) is sent, through resistor R5, to a Darlington pair consisting of Q2 and Q3. Since the encoder doesn't produce a signal intensity suitable for recognition by the central office (CO), this simple amplifier solves the problem. Its output is connected to the next section we will discuss, affectionately called the audio line interface.

The line interface incorporates metal-oxide varistor MOV1, capacitor C2, transformer T1 and diodes D6 and D7. The purpose of this stage is to match the electronic circuitry with the standard 600-ohm telephone line impedance. In addition to better sound quality, this section also insures the telephone will "seize" the line by presenting the proper impedance when off-hook.

Line compatibility is accomplished by 600-ohm-to-600-ohm matching transformer T1, while capacitor C2 blocks DC. MOV1, rated at 130 volts, acts as a surge suppressor for lighting-induced voltage spikes, and zener diodes D6 and D7 clamp the 90-volt ringer signal to about 3.9 volts.

Now, let's look at the microphone, or transmitter circuit. The mic is of the "electret" variety, and is part of the headset. Since the most commonly available microphone headsets are manufactured for computer use, and utilize the three-wire style microphones (a separate plus "+" lead for the positive voltage), I incorporated a three-lead mic in this circuit.

As a sidebar, electret mics are comprised of a capacitor where one plate acts as the diaphragm. When sound vibrations hit this plate, the capacitance is changed relative to the sound. This change is then amplified by a field-effect transistor (FET), and that is the reason the mic needs a source of power.

The microphone output is sent to a two-transistor amplifier stage that is capacitively coupled at both ends by C3 and C4. Resistor R8 sets the output impedance at about 1,000 ohms which matches the input of the speaker circuit. Additionally, the amp output is sent to the audio line interface for transmission to the telephone system's central office.

Some of the mic output is directed to the speaker or "receiver" section as sidetone, so zener diode D8 is included to "clip" the input to this section. Ideally, sidetone should be loud enough to be easily heard over the speaker, but not so loud as to drown out the party you are connected with.

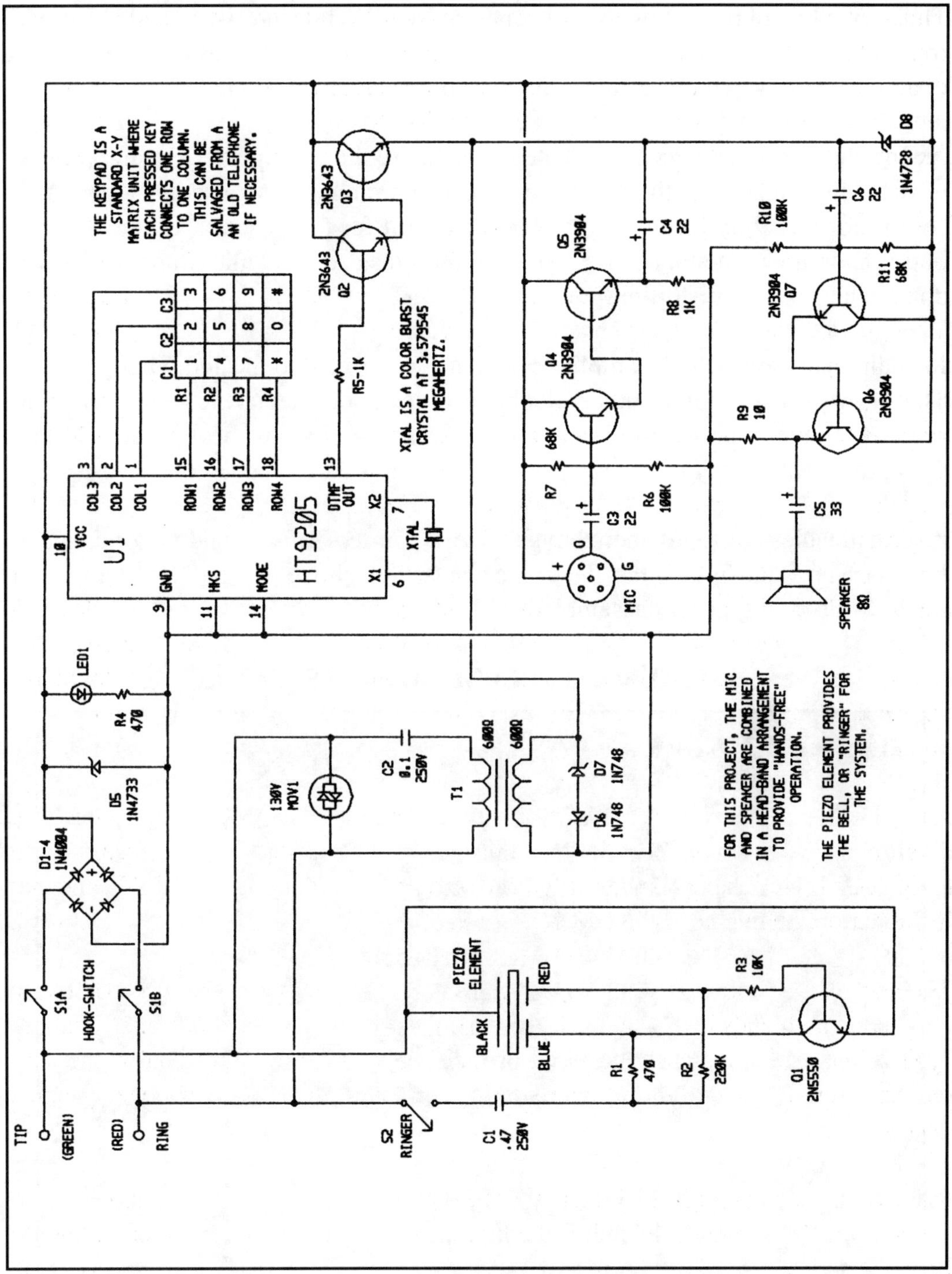


**Figure 9-1.** The hands-free telephone.

Thus, a method of regulating the sidetone is needed. While a resistor could be used, I found the zener diode did a far better job. Using a 1N4728 reduces the input voltage to about 3.3 volts, which produces an appropriate sidetone level.

As for the receiver, this is a duplicate of the microphone amplifier circuit. Well, almost a duplicate. Again, the basic two-transistor arrangement is used, but the impedance resistor (R9) is 10 ohms to adjust the output for the 8-ohm speaker. Also, the output coupling capacitor (C5) is 33 microfarads instead of 22 microfarads. This was done to help reduce distortion.

In addition to diode D8, the amplifier's input is connected to the audio line interface. This allows sound from the other telephone, or telephone device, to be heard on your receiver. While I admit being a master at stating the obvious, beyond that, I'm not going to elaborate. I think the receiver's operation is more than evident.

And that, ladies and gentlemen, boys and girls, misters, misses and mizes, and anybody else listening, is how the electronic headset telephone works. Now, let's examine how the prototype was assembled.

## HEADSET TELEPHONE PARTS LIST

| SEMICONDUCTORS: | |
|---|---|
| U1 | HT9205A DTMF Tone Generator (Digi-Key) |
| Q1 | 2N5550 NPN Transistor |
| Q2, 3 | 2N3643 NPN Transistors |
| Q4-7 | 2N3904 NPN Transistors |
| D1-4 | 1N4004 400V Rectifier Diodes |
| D5 | 1N4733 5.1V Zener Diode |
| D6, 7 | 1N748 3.9V Zener Diodes |
| D8 | 1N4728 3.3V Zener Diode |
| LED1 | Red T-1 Miniature Light-Emitting Diode |
| RESISTORS: | |
| R1, 4 | 470 Ohm 1/4W Resistor |
| R2 | 220,000 Ohm 1/4W Resistor |

(Continued Next Page)

| | |
|---|---|
| R3 | 10,000 Ohm 1/4W Resistor |
| R5, 8 | 1,000 Ohm 1/4W Resistors |
| R6, 10 | 100,000 Ohm 1/4W Resistors |
| R7, 11 | 68,000 Ohm 1/4W Resistors |
| R9 | 10 Ohm 1/4W Resistor |
| CAPACITORS: | |
| C1 | 0.47 µF 250V Mylar Capacitor |
| C2 | 0.1 µF 250V Mylar Capacitor |
| C3, 4, 6 | 22 µF Electrolytic Capacitors |
| C5 | 33 µF Electrolytic Capacitor |
| OTHER COMPONENTS: | |
| S1 | DPDT Slide Switch |
| S2 | SPST Slide Switch |
| MOV1 | 130V Metal-Oxide Varistor |
| MIC | 3-Lead Electret Microphone (part of headset) |
| SPKR | 8 Ohm Speaker (part of headset) |
| XTAL | 3.579545 MHz "Color Burst" Crystal |
| Piezo | 3-Lead Piezo "Buzzer" Element |
| T1 | 600-to-600 Ohm Matching Transformer (Jameco: 141621) |
| KEYPAD | 12-Key, "X-Y" Matrix Standard Telephone Keyboard (These are available all over the surplus market) |

MISCELLANEOUS:

PCB Materials, Hardware, Solder, 4 1/2" x 2 3/4" x 1" Plastic Case (Radio Shack), Telephone Line Cord, Dry Transfer Lettering, 18-Pin IC Socket, LED Holder, Etc.

## CONSTRUCTION

As with most electronic projects, either printed-circuit board (PCB) or point-to-point wiring can be employed. And, as usual, I recommend the PCB method. To that end, I

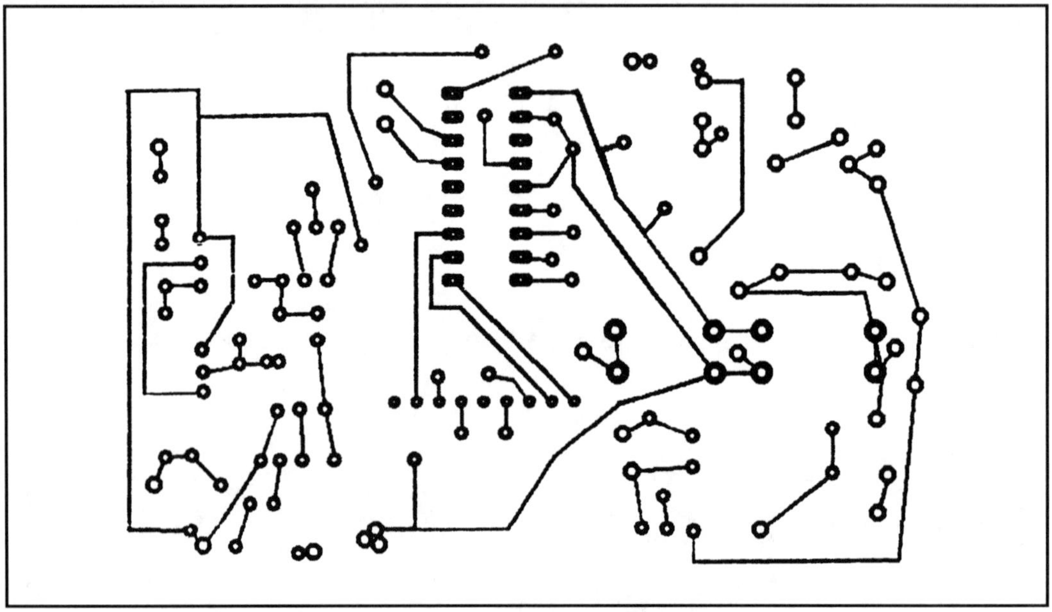

**Figure 9-2.** Headset telephone PCB foil pattern.

have included *Figure 9-2*, the foil or solder-side pattern for making a PCB, and *Figure 9-3* that shows the parts placement for said board.

Since I wanted the phone to be small and portable, I chose a 4 1/2 inch (114 mm) by 2 3/4 inch (70 mm) by 1 inch (25 mm) case, and the associated PCB ended up 4 inches (101 mm) by 2 inches (50 mm). This size accommodated all parts, except the switches, plugs and LED, leaving a little room to spare.

With the PCB in hand, start by installing the jumpers and IC socket, then the resistors, diodes and capacitors. Also, if the keypad comes with a header, now is the time to place it. Follow this with the transformer, transistors and crystal. That helps protect the delicate and heat-sensitive components. Don't install U1 in its socket just yet.

Now, solder appropriate lengths of hookup wire to the LED, switches and jacks, then connect the leads to the board. The piezo element is next, followed by the telephone line cord.

With all that done, it is time to make a quick voltage check. Plug the line cord into a modular jack, close hook switch S1 and check for 5.1 volts (or something very close)

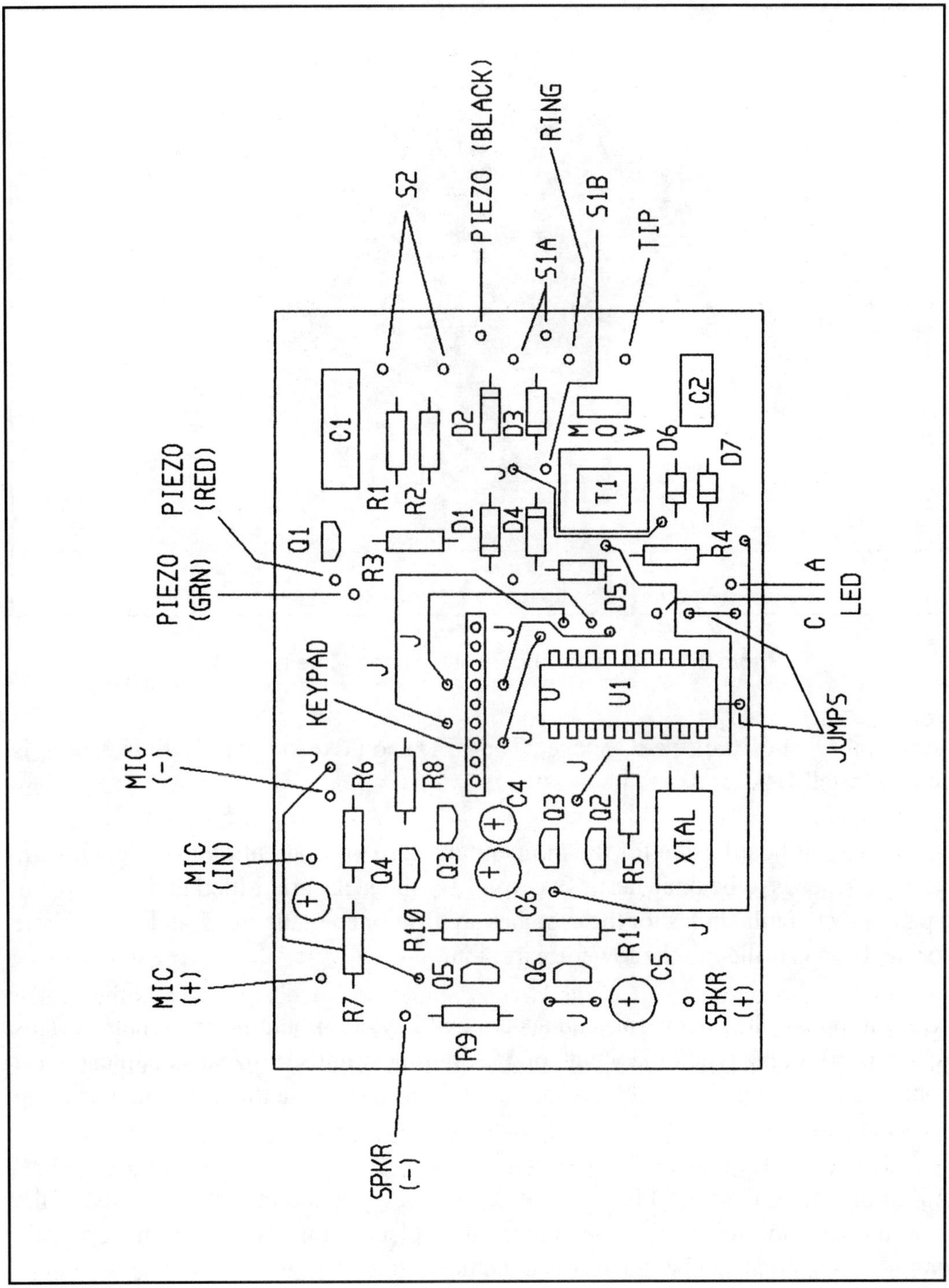


**Figure 9-3.** Headset phone PCB parts placement.

**Photo 9-1.** The stuffed board of the hands-free telephone.

across pins 9 and 10 of the IC socket. The LED should also light. If all is well, it is safe to install U1.

Set the circuit board aside for the moment and turn your attention to the enclosure. Very little needs to be done here. Two holes for the jacks are drilled in the end plate, a pair of rectangular holes for the switches are fashioned in the sides, and a small hole for the LED is drilled in the lower top region.

Also on top, arrangements for lead access and keypad mounting are required. This will depend on the type of keypad you use, but most units utilize a ribbon cable for connection between the switch contacts and PCB. Usually, a thin elongated opening does the job.

As for mounting the pad, if it comes with hardware you are in luck. But most of the time the keypad will need to be cemented in place. Some, however, have plastic mounting pins that can be glued or melted down to fasten the assembly.

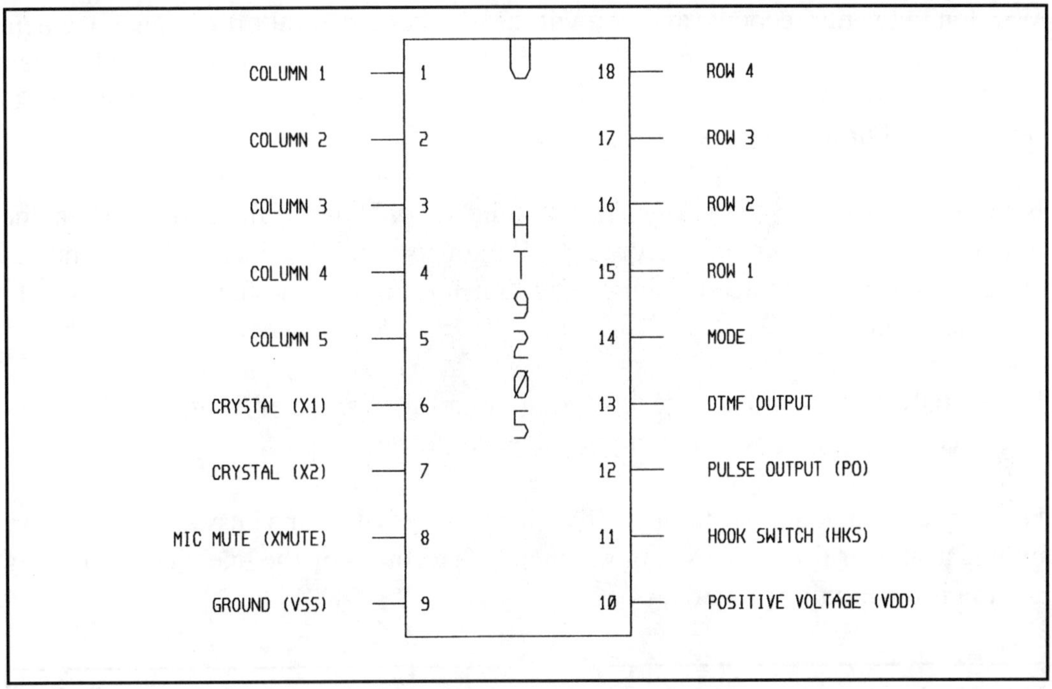


**Figure 9-4.** HT9205A pinouts.

Last, a small hole for the line cord needs to be drilled at the top end of the case. It is a very good idea to tie a knot in the line cord, or use some other form of "strain relief" to prevent the connections from breaking loose. Remember, there will be a reasonable amount of strain on the cord.

With everything in place, plug the keypad cable into the header, or solder it to the PCB, and it's time to test the circuit. No internal adjustments need to be made, but it might be wise to confirm proper operation before closing up the case. It's possible you might still need access to the board.

## CIRCUIT TESTING

Testing the system is a snap. Put switch S1 in the "on-hook" (open) position and plug in the headset and line cord. Flip S1 to the "off-hook" (closed) position, and you should hear a dial tone in the headset. Also, LED1 should be glowing.

Wait ten to twenty seconds, and you will hear either a central office announcement telling you to hang up, or the well-known attention-getting beeping tones that alert you to an off-hook condition. If this happens, that's good! It's a signal that the CO recognizes your telephone is on the line.

Now, press any button on the keypad. Two things should occur: first, you will hear the tone pair in the headset; and second, the dial tone should disappear. The second occurrence indicates the encoder signal is sufficient to activate the central office switching equipment.

Incidentally, you should be able to hear, in the speaker, some of whatever the microphone is picking up. This, of course, is sidetone feedback.

Next, try dialing your own number. If all is well, you will hear a busy signal. Now, try dialing a number like local time or weather. If that comes on the line, then the dialing section is functioning properly.

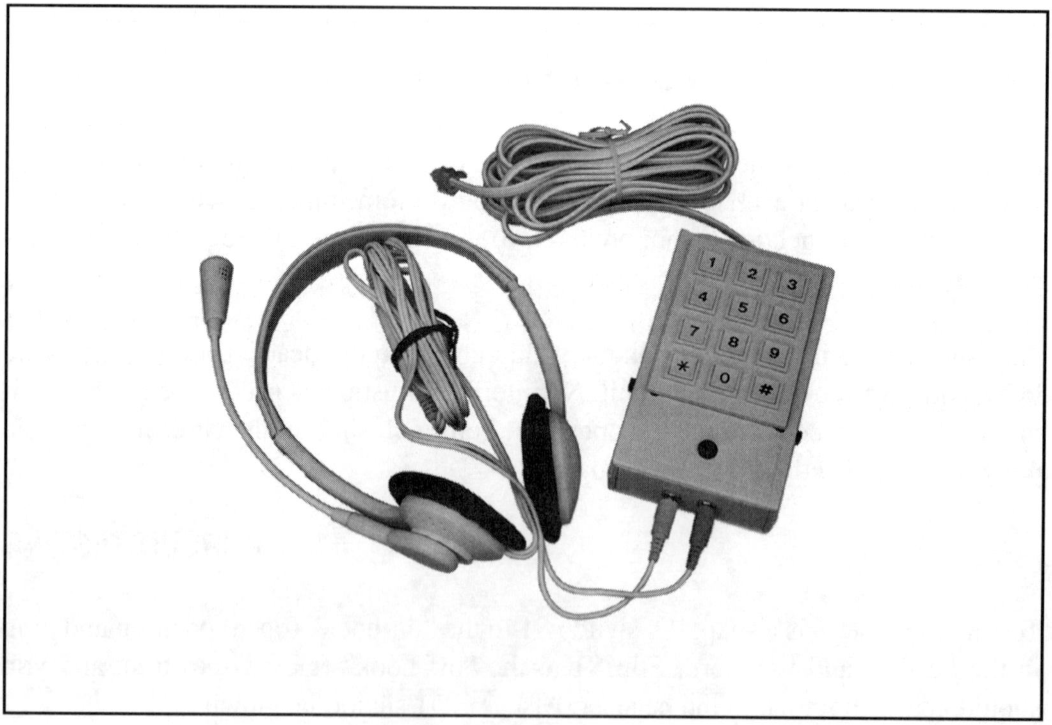


**Photo 9-2.** The complete hands-free headset telephone with headset and keypad control box.

Last, try calling a friend, relative, enemy, some warm body and talk with them. You should be able to carry on a normal phone conversation with your party, but do ask about the sound quality on their end. Expressly, if it sounds normal.

And, as I said, testing is a snap. When you have performed all the above, you have checked the phone, and it's ready to use. Normally at this point I would include an "Operations" section, but I don't want to get ridiculous. I mean, my mother didn't raise no fool! I've got to assume all of you already know how to operate a telephone.

## CONCLUSION

Voila! You have constructed your very own headset, hands-free telephone. Congratulations! I know, I sound like the first page of an instruction manual. "Congratulations! You're now the proud owner of this fine product," etc., etc. But this time I think you are going to like your fine product very much.

At the outset, I remarked that a headset phone might appear to be more of a luxury than a necessity. In all honesty, it's rarely a necessity, but it sure can be handy. Anytime you need your hands free while on the phone, this device is made to order.

As for long conversations, it is mighty nice not to have to hold that receiver up to your ear. Just slip the headset on and relax during the call. And, of course, there is the initial reason I mentioned regarding keyboards and telephones.

On a last and perhaps more pragmatic note, building this device goes to the heart of this book; that is, to help take the mystique out of telephones and the system that supports them. All projects aside, if this text accomplishes that alone, it's doing its job. Naturally, the projects are nice.

Have fun with this one! It is a great endeavor!

# CHAPTER 10

# THE LONG-DISTANCE CALL RESTRICTOR

Do you have an employee who won't follow the NO LONG-DISTANCE CALLS rule? Do you have children who love to dial every toll number they can get their hands on? Do you have a telephone bill that is about to drive you to chapter 11? Well, if so, do I have the gadget for you!

Here is a simple device that detects attempts to dial long-distance phone numbers, and when it senses a culprit, momentarily disconnects the telephone line. The phone connection comes right back, but a second, third, forth attempt, etc., will result in the same consequence—NO CONNECTION!

How does it perform its miraculous feat? Patience, I'll get to that later. But, for the moment, let me hawk this one a little further. We are talking about a handful of *cheeeeap* parts, and a couple of nights' work (maybe a weekend) for all this peace of mind (not to mention the POWER it will bring you).

With this project completed and installed, your seemingly endless warnings regarding telephone abuse will no longer go unheeded. Now when you lay down the "No Long-Distance Calls" law, even if certain parties want to violate it, they simply can't.

Sound good? Read on!

## THEORY

The principles behind our phone restrictor are as follows: first, when you lift the telephone receiver, a dial tone is present until you dial the first digit; second, each long-distance number starts with a "1"; and last, for a slight moment, the dial tone and first-dialed number remain on the line simultaneously. So slight, in fact, that you probably cannot detect this condition. However, as we all know, electronics is QUICK; thus, it *does* see both signals.

With these thoughts in mind, the rest should be easy. Ah, sure it is! All we need is a circuit that recognizes the presence of both the dial tone and digit "1" Dual-Tone Multifrequency (DTMF) tone pairs. Add some logic to control a relay, and you've got it.

Actually, since the dial tone is unique unto itself—that is, neither the 350-hertz or 440-hertz tones are shared with anything else—we only need to detect one of those frequencies, which makes things somewhat simpler. By employing three LM567 phase-locked loop tone detectors (decoders), the 697-hertz and 1209-hertz pair for "1" and either of the dial tone components can be discerned. The IC outputs are then used to control the logic section.

Let's take a look at *Figure 10-1*, and delve into this scheme in more detail. Modular jack J1 provides connection to the telephone line. Note that the "ring" lead is routed to the contacts of relay RLY1, and the "tip" rides the circuits ground rail. Also note the "ring" furnishes the circuit with an input signal through resistor R1.

The other side of R1 is connected to the input of each tone detector (U1, U2 and U3) through coupling capacitors C3, C7 and C11. This brings the DTMF tones to the decoders. When tuned properly, each IC can zero in on a rather narrow frequency.

Since all LM567s are wired identically, we'll examine just the first decoder circuit, U1 and company. Pins 1 and 2 handle the output (O) and low-pass (LO) filter capacitors C4 and C5 respectively. Pin 3 is the input line, while pins 4 (VCC) and 7 (Ground) manage the power.

Tuning, or "timing," is controlled by pins 5 (Resistor) and 6 (Capacitor) using the R/C tank of R2 and C6. Finally, pin 8 is the output which goes "low" when the desired frequency is detected. Light-emitting diode LED1 and dropping resistor R3 act as a visual indicator of this detection, with the LED's anode tied to the positive rail. Using the values signified, each phase-locked loop can cover the entire range of telephone system audio tones. I will cover the tuning process in a later section. So for now, let's move on to the logic.

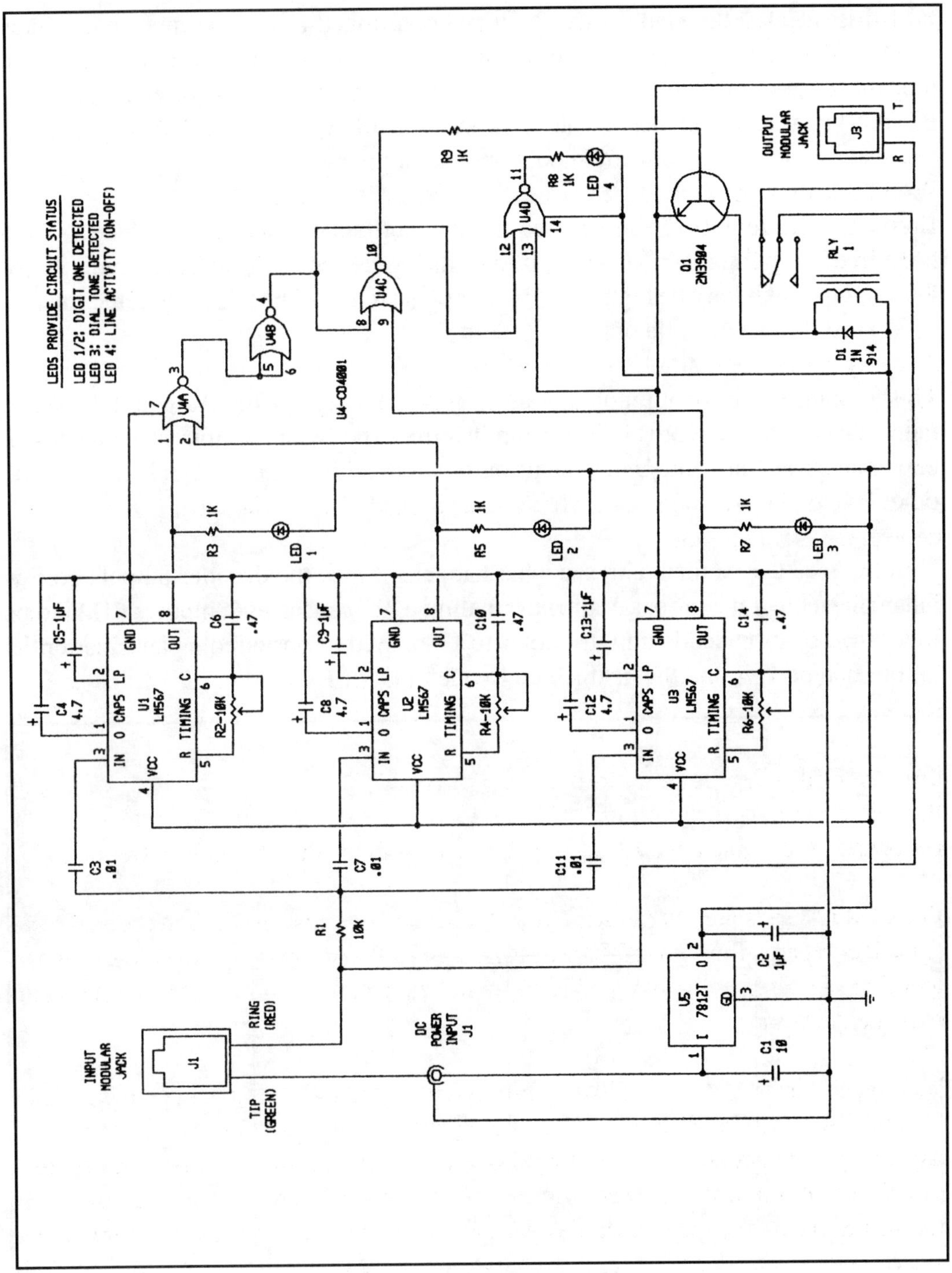


**Figure 10-1.** Long-distance call restrictor.

IC U4 is a CD4001 Quad 2-input NOR gate arranged as a three-input NOR gate. Actually, only three gates are used for this, with the fourth, U4d, employed to light the "line active" indicator. By the way, a complementary metal-oxide semiconductor (CMOS) IC was chosen here because the system voltage is 12 VDC. A transistor-to-transistor logic (TTL) chip could not handle that potential.

Looking at *Figure 10-1*, you will see that the outputs of the digit "1" detectors are connected to the inputs of U4a. Applying the NOR gate truth table (*Figure 10-2*), U4a's output will only be high when both inputs are low. Thus, it takes detection of a "1" to result in a high U4a output.

This output is tied to both inputs of the second NOR gate U4b. As with all "NOT" gates, this effectively turns U4b into an inverter. The second output is connected to one input of the third gate U4c, and the output of the third decoder, U3, is wired to the other U4c input.

Now, let's take a moment and see what happens. Since the decoder outputs go low when they detect the proper frequency, dialing a "1" causes both inputs of U4a to be low. That, in turn, sends a high output to U4b. With U4b wired as an inverter, its output will be low, and that is directed to one input of U4c.

NOR GATE
LOGIC
TRUTH TABLE

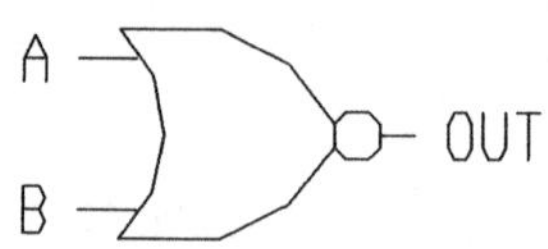


| INPUTS A | B | OUTPUT |
|---|---|---|
| LOW | LOW | HIGH |
| LOW | HIGH | LOW |
| HIGH | LOW | LOW |
| HIGH | HIGH | LOW |

**Figure 10-2.** NOR gate logic truth table.

With U3 tuned to the dial tone, whenever the phone is taken off-hook, U4c's second input will be low. Thus, when a "1" is dialed as the first digit, for an brief instance, all three detectors have low outputs. This results in U4c's output swinging high.

Since that output is tied to the relay, through driver circuit transistor Q1 and resistor R9, RLY1 is activated, and the "ring" line is disconnected. This momentarily cancels the attempted call, and returns a dial tone to the phone. Furthermore, any additional efforts to dial long distance will meet the same discouraging fate. In this configuration, Q1 is utilized as a switch. When power is applied to the base, the positive potential on the relay coil is allowed to pass through the transistor to ground. Current moving through the relay's inductor sets up a magnetic field that results in the switch contacts changing state (in this case from normally closed to normally open). That action breaks the ring line connection to the output modular jack.

Note that a small signal diode is wired in reverse across the relay coil. This is done to prevent the coil's collapsing magnetic field, which occurs when power is removed, from generating a voltage spike that could damage Q1.

By using two tone pair sets to activate the relay, the "1" can be dialed as part of a local number without interrupting service. This is because the central office connection will only be broken when both a dial tone and a "1" are present on the line. And personally, I've never run into a local exchange that began with the digit "1". That seems reserved for specialized services, such as long distance, emergency and directory assistance.

As stated, the last NOR gate (U4d) is used to indicate "active" telephone line status; that is, whenever LED4 is on, the phone line is in service. This is done by tying one U4d input to U4b's output, and the second input to ground. U4d's output is tied to Vcc through the LED4/R8 network, thus the output has to be low for LED4 to shine. However, if U4b's output is low, this configuration causes U4d's output to swing high, and the diode goes out.

Again, applying the truth table, it is seen that a NOR gate output is always low unless both inputs are low. Then, the output goes high. With our arrangement, one input will be high and the other low unless a "1" and a dial tone are detected. That condition puts a low on both inputs, with a resultant high on the output which extinguishes the indicator. Thus, LED4 stays on until you try to dial a long-distance phone number.

## LONG-DISTANCE CALL RESTRICTOR PARTS LIST

SEMICONDUCTORS:

| | |
|---|---|
| U1, 2, 3 | LM567 Tone Decoder Phase-Locked Loops |
| U4 | CD4001 Quad NOR Gate |
| U5 | 7812T Positive 12V Regulator |
| D1 | 1N914 or 1N4148 Signal Diode |
| Q1 | 2N3904 NPN Transistor |
| LED1-4 | T-1 3/4 Jumbo Light-Emitting Diodes (any color) |

RESISTORS:

| | |
|---|---|
| R1 | 10,000 Ohm 1/4W Resistor |
| R2, 4, 6 | 10,000 Ohm, 10-Turn PCB Potentiometers |
| R3, 5, 7, 8-9 | 1,000 Ohm 1/4W Resistors |

CAPACITORS:

| | |
|---|---|
| C1 | 10 µF Electrolytic Capacitor |
| C2, 5, 9, 13 | 1 µF Electrolytic Capacitors |
| C3, 7, 11 | 0.01 µF Ceramic Disk Capacitors |
| C4, 8, 12 | 4.7 µF Electrolytic Capacitors |
| C6, 10, 14 | 0.47 µF Mylar Capacitors |

OTHER COMPONENTS:

| | |
|---|---|
| J1, 3 | RJ11 4-Pin Modular Telephone Jacks |
| J2 | Coaxial Power Jack (matches DC power plug) |
| RLY1 | 12V Relay w/SPDT Contacts (see text) |

MISCELLANEOUS:

PCB Materials, Solder, Hardware, IC Sockets, Hookup Wire, Case (if desired), 14 VDC Wall Transformer, Etc.

In reality, U4d is detecting the presence of only the digit "1" tones. This is because its alternating input (the other input is hardwired to ground) is tied to U4b's output. However, for our purposes, that'll work!

If all this seems confusing, don't feel bad. Logic always is. At least for me, anyway. In fact, sometimes, it seem down right illogical. However, if you follow the truth table through the entire progression, it will become clear. It may take a couple of passes, but eventually the "logic" bears out.

OK, so there it is. Fairly simple technology that accomplishes a needed goal. Hopefully, this is what electronics is all about—although, I will admit the technology isn't always simple, or at least this simple. With those gems of wisdom stated, let's move on to building this varmint.

## CONSTRUCTION

I always like to keep projects as reasonably small as possible, but with this one, size is not a dictating factor. You probably are not going to try to install this inside a telephone, so a little bit larger board is acceptable here.

Hence, I went with a 2 1/2 inch (64 mm) by 5 inch (127 mm) printed-circuit board. This provided a good compromise between size and convenience. *Figure 10-3* furnishes the foil or solder-side pattern, while *Figure 10-4* is the parts placement. Of course, you could go the "point-to-point" route if you desire.

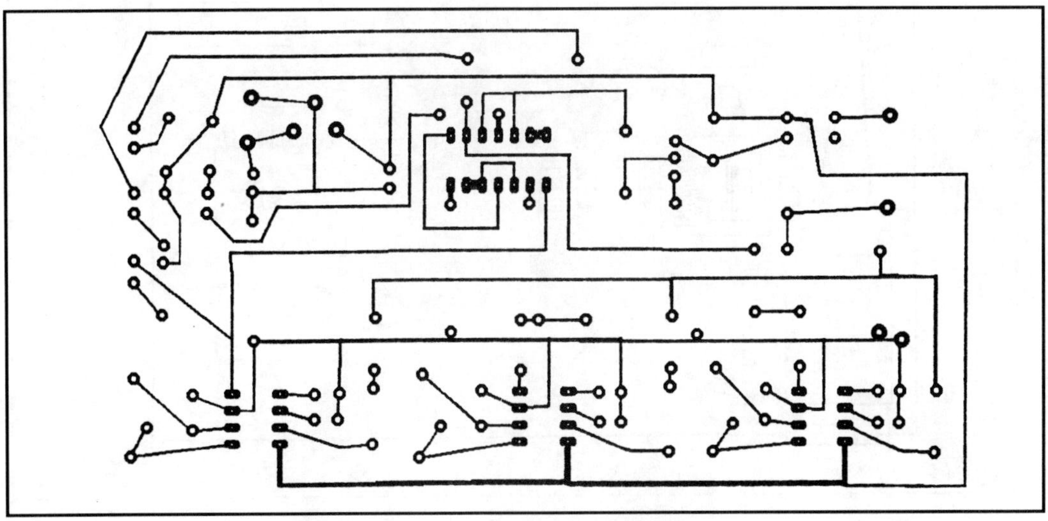

**Figure 10-3.** Call restrictor PCB foil pattern.

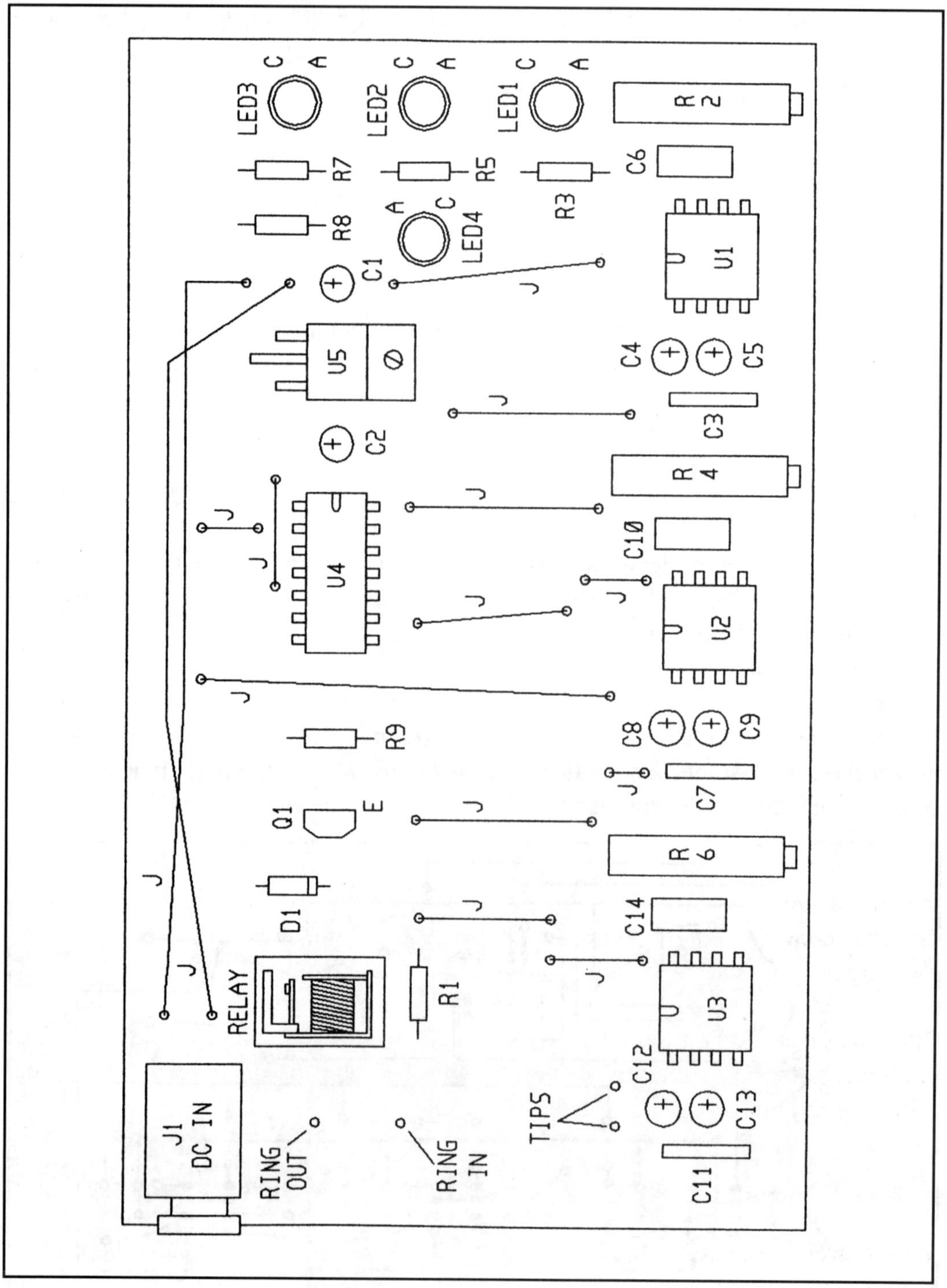


**Figure 10-4.** PCB parts placement.

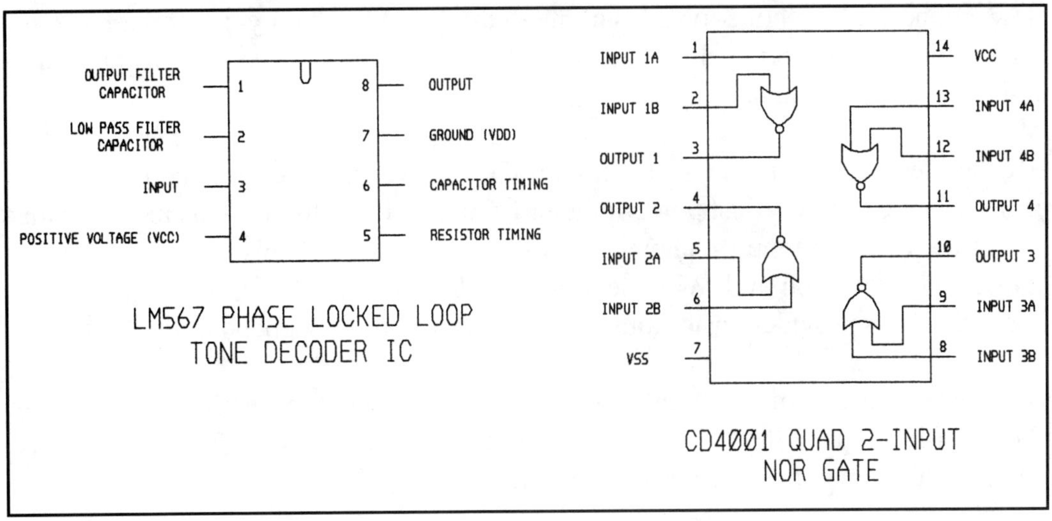


**Figure 10-5.** LM567 and CD4001 pinouts.

Using the PCB, start with the jumpers and integrated circuit (IC) and relay sockets. Follow this with the resistors, capacitors, LEDs, diode, transistor and regulator. Finally come the hookup leads for the power and modular jacks. Lead length will depend on how you plan to install the unit, or what you plan to install it in.

Once that is done, connect the DC power source to the power jack, and test the voltage outputs at the following locations: U5 at pin 2; U1, U2 and U3 at pin 4; and U4 at pin 14. In each case, the measurement is taken between the designated pin and ground. Additionally, the power supply will need to provide at least 14 VDC.

If you receive a 12-volt reading at each point, then all is well—at least, as far as the power is concerned. Disconnect the DC source, and it is now safe to install the remaining four ICs and relay in their sockets.

That takes care of the assembly. If you plan to enclose the unit in a case or piece of equipment, it is best to complete the following Testing and Tuning section first. That will get everything ready for proper operation.

## TESTING AND TUNING

With power reconnected, hook the "input" modular jack (J1) to your telephone line with a suitable length of cable. At this point, LED4 should light. That, of course indicates the phone connection is in service.

Now, connect a telephone to “output” modular jack J2. When you lift the receiver, you should hear the normal dial tone. If not, be sure the relay is properly installed, and in the “inactivated” state.

OK, here comes the tuning part. Pick up the receiver and be sure the dial tone is present. Now, slowly adjust potentiometer R6 until LED 3 lights. This indicates that one of the two dial-tone frequencies has been detected. For this section, it doesn’t matter which tone you find. With the prototype, I used multiturn PCB potentiometers. Their precision provides easier tuning, as well as improved dependability.

With that done, it’s time to set the two remaining decoders for the digit “1.” With the phone off-hook, press the “1” and slowly rotate potentiometer R2. When LED1 lights, U1 is tuned to one of the two proper tones.

Since both 697 and 1209 hertz are shared with other digits on the phone dial, press the “2” digit and see if LED1 lights again. If so, you have found the 697-hertz (row 1) frequency. If not, 1209 hertz (column 1) is tuned. You might try pressing digit “4” just to be sure. LED1 illuminating will confirm 1209 hertz. Either way, the next decoder has to be tuned to the other tone.

Again pressing digit “1,” adjust potentiometer R4 until LED2 lights. Again, press digit “2” and if LED1 lit before, you don’t want LED2 to light. If it does, you have both decoders tuned to the same tone. So, with “1” activated, continue to adjust R4 until LED2 lights a second time. Now, press digit “4” and LED2 should glow. That will mean you have U1 and U2 each tuned to a different digit “1” frequency.

Once you have all the LM567s tuned, pick up the receiver and LED4 will illuminate. Press “1” and LED4 will momentarily go out, and you should hear a new dial tone when the diode resumes glowing. If all that happens, a couple of “atta boys” are in order, as the restrictor is calibrated and ready for service.

## APPLICATION

The previous section described operation of the device, and I think that gives you a pretty good idea of how the system protects your telephone line. So, rather than elaborate on that aspect, let me discuss installation of this unit for optimum performance and maximum security against tampering.

Fortunately, once the restrictor is calibrated, it will rarely, if at all, need additional attention. So, this allows you to place the device in an obscure location. That will help keep it away from those who might want to sabotage your efforts.

Of course, if you plan to take the system on- and offline, it will need to be accessible, at least to you. Naturally, there are going to be times when you want or need to make long-distance calls. For that, the restrictor will need to be disabled. Also, "800" and other "WATS" (Wide Area Telephone Service) type numbers will be blocked by this device, as well as "1-911" and "1-411."

One easy way to cancel the unit is a good old-fashioned "ON-OFF" switch. And, to help prevent unauthorized "ONing" and "OFFing," a "key" style electrical switch, such as those used on desktop computers, is highly recommended. Unless you are dealing with someone good at picking locks, a key switch should keep the system reasonably secure.

Another area of peril is the power source. Disconnecting the power has the same effect as turning the switch off. So, if you use an enclosure, it's advisable to keep the power jack inside, and out of reach. That should prevent some unscrupulous rogue from coming along and unplugging your telephone guardian.

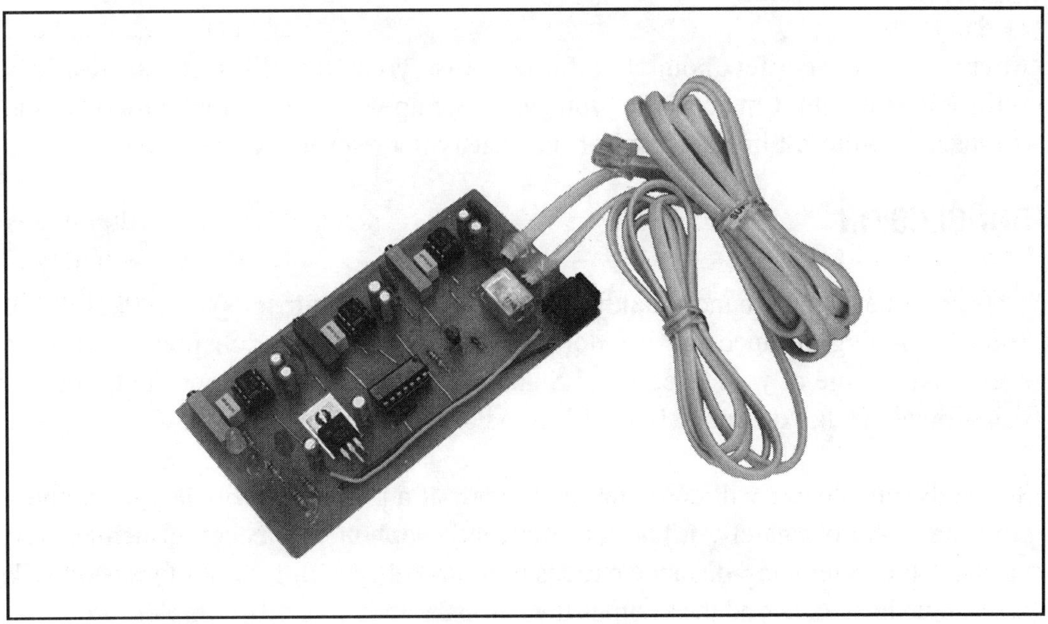

**Photo 10-1.** The completed project can be housed in a case or used "as-is."

The same logic applies to the modular jacks. If the line cable and phone connections are easily accessible, an "in-line" coupler could be used to bypass the unit. That, needless to say, defeats the purpose of installing the restrictor.

As for locations, one of the best is wherever the "entrance bridge" or telephone terminal block is situated. The entrance bridge is a small rectangular box, usually with a snap lid, that provides a terminal strip for the incoming telephone line.

These are often found in closets that access the central air/heating system, near interior fuse boxes or in other maintenance-oriented areas. It is from here that all extension lines originate—thus, an ideal place to position our device.

The incoming line can be wired directly to the restrictor, with extension lines connected to the output modular jack. In that fashion, whenever the restrictor is operational, all telephones on the line are affected. Naturally, the key switch, if used, controls system functionality.

Of course, the device can be placed on any individual phone that is susceptible to long-distance abuse. Again, proper security measures will be necessary to prevent interference. But, with a little ingenuity, that usually can be accomplished. Sometimes, even clandestinely.

Anyway, these are a few pointers on how to use your long-distance call restrictor. With some thought, I'm confident you can come up with many more. Often, a "sudden need" will result in a new and/or previously not envisioned application.

## CONCLUSION

There you have it! A handy-dandy method to protect your telephone (and pocket book) from long-distance use. So, don't file that chapter 11, or chapter 13, or whatever. Take charge of your telephone! Stand up to the problem! Don't let them push you around! Well, you know. It sounds good on paper.

Seriously, this device will come in quite handy if a problem exists in your home or workplace. Unfortunately, at least at times, the telephone has become such a part of our lives that often long-distance expenses go unrealized. But, even those short talks can add up in a hurry, and then comes the coronary-inducing bill, complete with short breath and chest pains. Ouch!

So, there are a number of excellent reasons to build this gadget. Aside from the cost and challenge to your authority, there is also knowledge to be gained. The technology is not new, but this device illustrates how to put that technology to use. And, that makes a statement in defense of the older "industry standard" ICs. I grow weary of hearing how those old chips are obsolete.

Too often these days, the faithful chips that have served us for so many years are drowned in the wave of newcomers. I'm all in favor of new integrated circuits. They bring exciting avenues of exploration and utility, but I like to retain soft spots in my heart for some of those ICs I grew up with. After all, if it had not been for the likes of the 555, 741, 339, 567 and others, where would integrated circuits, and the rest of us, be today? See there, I don't care what Natalie says. I remember my old friends. Just kidding, Natalie!

# CHAPTER 11

# HANDY COMMERCIAL PRODUCTS

Ever since the Federal Communications Commission's (FCC) 1979 ruling allowing connection of privately owned equipment to the telephone system, the marketplace has been flooded with commercial gear. Everything from multiline, multifunction telephones to devices that control your home appliances have appeared, and some of them are right interesting.

First off, though, let me state that you can't just attach anything you want to the telephone lines. Privately owned apparatus cannot damage telephone company equipment, or anyone else's, it cannot cause interference or degrade line quality and it must be safe for the user and telco employees. In other words, it has to conform to FCC regulations Part 68.

Actually, FCC Part 68 began life as mere guidelines drafted in 1976. This led, however, to a full set of stipulations that today runs several hundred pages in length. Of course, American Telephone and Telegraph (AT&T) wasn't very happy with the thought of outsiders being able to connect to their lines. As a matter of fact, AT&T's concern resulted in a lawsuit that eventually ended up in the U.S. Supreme Court.

However, they lost, and today an estimated 200,000 telephone-related products have come and gone, sometimes come again and sometimes stayed with us. So, let's take a look at a variety of these products. What will be presented here is by no means everything, as it is hard to keep up with all the new stuff. But hopefully you will get an idea of what is out there, what you might need and maybe what is just around the corner.

## NEW FACES ON OLD FRIENDS

One very attractive facet of the 1979 FCC ruling was the "face-lift" it gave the telephone. While I'm not quite old enough to remember the phones where you had to hold the speaker part up to your ear and yell at a horn-shaped microphone, I do remember the good ol' (and drab) black job that sat on the desk, or hung on the kitchen wall. It had a rotary dial, and all the personality of a lump of dirt. Sorry Ma Bell, but it's the truth, and you know it.

An indication they knew it was the 1950s and '60s public relations effort to revitalize the telephone's image. Those decades saw the arrival of the "Princess" phone, a stand-alone modernistic design that had a problem with falling over, and (THIS WAS THE BIG ONE) *colors*. Yes sir, for the first time, you could get a home phone in something besides black (shades of Henry Ford).

**Photo 11-1.** A typical Bell rotary dial phone of the late '50s/early '60s. This particular model had the "multi-line" switches at the bottom.

All of this was an improvement, but when you stripped away the glitter, you still had a microphone, speaker and dialing mechanism. At the time it was functional, but no pizazz! However, the '79 decision would change all that. Now, telephones would have a little sex appeal, and the industry would never again be the same.

**Photo 11-2.** The ol' hang-on-the-wall, usually-in-the-kitchen, touch-tone telephone.

Today, you have a choice. Boy do you have a choice! There are phones in every color of the rainbow, shaped like cartoon characters, footballs and automobiles, just to mention a few. Some light up with neon, others crow like a chicken and others let you see inside.

You can get little numbers (keypads), big numbers, medium size numbers, strange shaped numbers. They still have the one that stands upright (and still wants to fall over), and wall phones. There is the "Slim-Fone," "Trim-Fone," "Mini-Fone," "Tudor," "Compact," "High Park," "North Park," "Seashell," "Contempra," "Diana" and who knows what else.

**Photo 11-3.** A typical desk style telephone with 10-memory dialing, hold and flash.

**Photo 11-4.** A small desk phone with redial, hold and speaker functions.

The desk versions now have built-in speed dialers, number memories, caller ID units and answering machines. Others come with microphone/headphone assemblies that make them "hands-free." Then, of course, there's the wireless phone that lets you move around while you talk. And, let's not forget the generic handset (can replace most conventional receivers) that has the built-in micro tape recorder in case you want to document your conversation. The list goes on and on!

**Photo 11-5.** A full-function speakerphone, complete with memory dialing, auto redial, pause and hold.

Now, if it seems I'm being sarcastic or critical here, let me correct that impression. I am not! Quite the contrary! I think all of it is great. But, the variety can be overwhelming at times.

Naturally, the telephone itself has received the most attention. Style and variety are two elements many companies have targeted. Again, at the risk of sounding critical, the phone is a device that is hard to glamorize. But a trip to your local telephone specialty store will quickly reveal the lengths manufacturers have gone to do just that.

One key has been to keep up with the times; that is, change the form often, but stay within contemporary limits. If TVs, VCRs and other electronic equipment are adding curves to their designs, use the same approach for phones. This will provide a modernistic appearance without being radical.

Another deciding factor is utility. Large keypads for children and people with poor eyesight, lighted dials for bedroom use, and headsets for those who have to keep both hands free are examples of how companies weigh public needs and design their products accordingly.

**Photo 11-6.** A 'cell' style land-line telephone, complete with digital display to show number being dialed, and volume control.

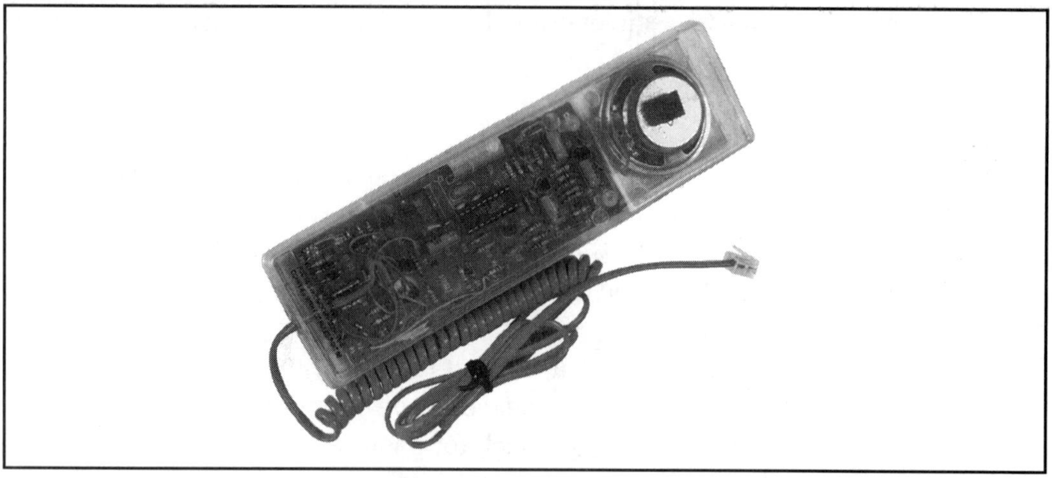

**Photo 11-7.** One of my favorites, this clear-case phone allows the user to see what's inside.

Memories for storing frequently-used numbers, displays that show dialed digits and other information, and wireless, or "cordless" phones, are a few more. Once a special need is recognized, it is definitely in the interest of manufacturers to try and meet that need. Thus, telephone variety is forever changing, usually for the better.

Some of my favorites are the cordless headset phone that clips to your belt and conveniently allows for movement and "hands-free" operation, the multifunction desk models with caller ID and memory dialing, and the "see-through" jobs. The tall one that's always tipping over? Nope, I'm not too fond of that one!

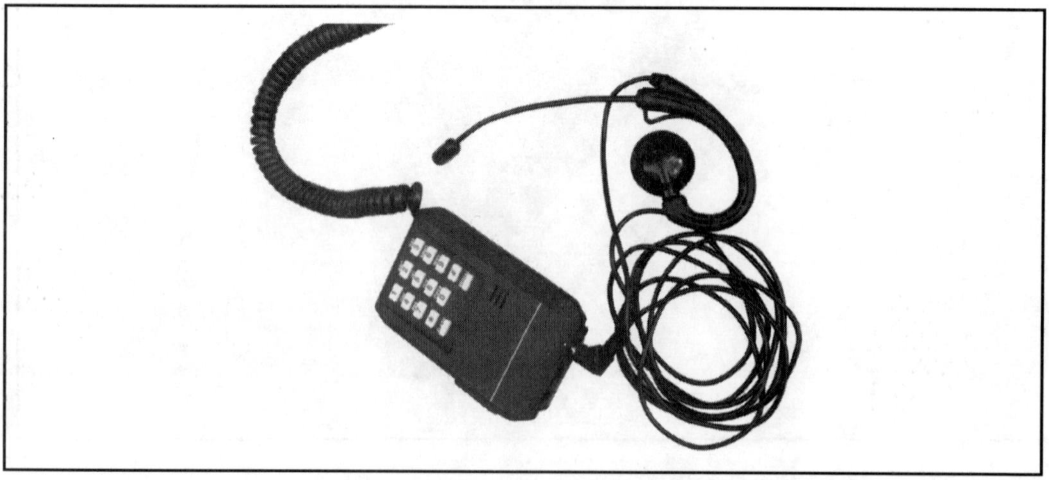

**Photo 11-8.** A small "belt-clip" headset style phone for "hands-free" operation.

Another area that has seen change for the better is speakerphones. The original models used a separate box that had to be fairly close to the person speaking. However, today's units, often built into desk phones, can hear you from anywhere in an average size room or office. This can be a very handy feature.

Along these same lines are the conference, or so-called "sound stations." These things are triangular in shape, with a speaker in the center and microphones at each point. Some also have a keypad on one side.

For conference style calls involving several individuals, they are hard to beat. Many models even include accessory remote microphones that plug into the main unit. This allows for even larger groups, and assures everyone is clearly heard.

Wireless phones have come a long way in the last ten years. With the original models, fading in and out and static were major problems—or, should I say annoyances. Today's cordless phones have virtually eliminated this nuisance either by scanning multiple frequencies to find a clear "channel" or employing the 900-megahertz band. Just below microwave, the 900 band provides very clear signal reception, and is rapidly becoming the frequency of choice for wireless operation.

**Photo 11-9.** The ever-popular "wireless" telephone. This unit communicates with a base unit that connects to the phone line.

Additionally, style and functionality is ever-improving. They now come in "flip-phone," belt clip and "cellular" designs, and many allow the handset to be used as a "roving" intercom. In this capacity, the base unit can alert and communicate with the handset.

Speaking of cellular, this is an area of almost continuous change. Many of the old standbys endure, while newer and smaller models hit the market on a regular basis. Some of the portables are so tiny they literally fit in a shirt pocket. This size does make the audio volume a little weak and the keypad rather small, but they do a remarkable job. I was amazed at what I recently saw in a cellular company merchandising store.

Most units accept external antennas, headsets and cables for data transfer from computers. With a laptop and a cellular phone you could truly run your business on the fly. Well, most of it anyway.

However, as astounding as these "analog" phones are, their "digital" cousins will really knock your socks off. (What an expression! I wonder where that came from? *Knock your socks off.* Anyway, the digitals are something to behold.)

**Photo 11-10.** Two cellular phones. A larger "bag" style system, and a small low-power handheld model.

In addition to digital clarity and performance, they offer such services as call waiting, call alert, call forwarding, voicemail and text/numerical paging. Many have built-in two-way radios and/or pagers, and with most services there are no roaming charges. Also, you can "pool" on-air time between several separate phones, and 2-line capability and 3-way calling is available.

Several companies are actively promoting digital cellular systems, and just as a guess, mind you, I expect to see this technology take off like a three-stage Saturn rocket. Of course, I could be wrong. It wouldn't be the first time.

So, in the future, if you don't know what to do for an afternoon, try visiting one of the many merchants that handle telephone stuff. Most discount stores, department stores and electronics outfits carry telephones these days. And, just about every mall has at least one shop dedicated to Bell's invention.

Also, don't forget your local cellular server. They too have a barrel full of monkeys just for your amazement and amusement!

**Photo 11-11.** An inside shot of a typical full-power mobile cellular telephone transceiver. This would be found permanently installed in a vehicle, or as a "bag" phone.

## THE OTHER STUFF

Probably the most prolific, as well as immediate, accessory to hit this market was the telephone "answering machine." The early machines were bulky, expensive and tended to be somewhat unreliable, at least regarding the magnetic tape section. The cassettes were notorious for jamming and even tangling.

When I refer to "early" machines, I'm talking about the post-1979 equipment. There were answering machines before that time, but they were owned and installed by Ma Bell.

However, it wasn't long before manufacturers got the message (excuse the pun), and the race was on. For example, the original "little black boxes" you had to carry around to remotely access your answering machine gave way to DTMF tone methods. This allowed you to check your messages with any touch-tone phone.

The next breakthrough came in the form of size. Employing microcassettes helped reduce machine bulk and also improved dependability. The smaller tape was less likely to "foul up," but it was still tape.

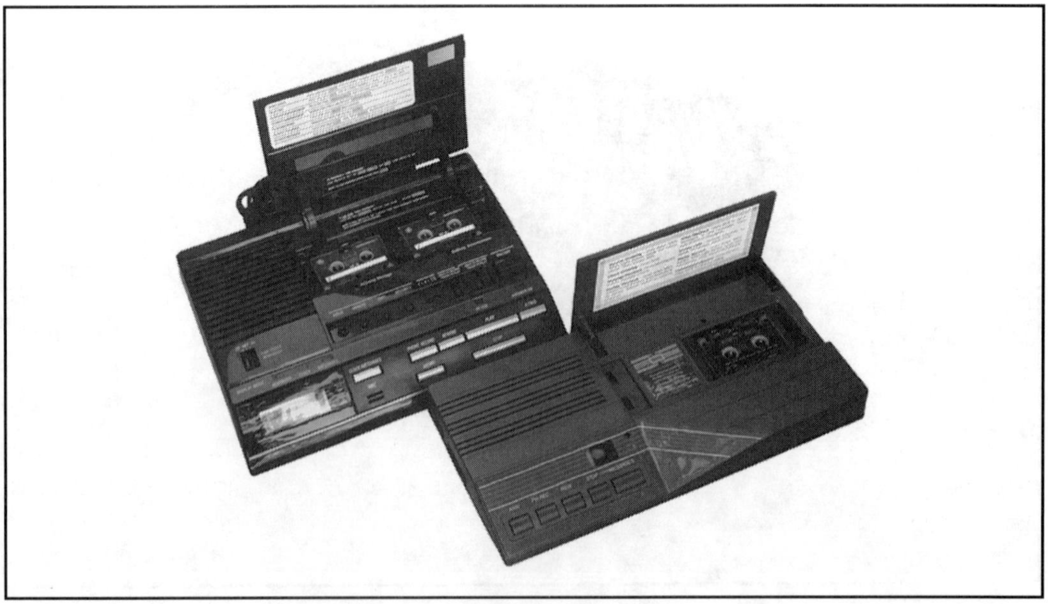

**Photo 11-12.** Two typical answering machines. The top unit uses two cassette tapes, while the bottom system has one tape and a digital memory for outgoing messages.

Thus, the next move took advantage of a new semiconductor technology—that was, storing voice messages on memory chips (ICs). The memory ICs, either battery backed-up dynamic random-access memory (DRAM) or electrically erasable and programmable read-only memory (EEPROM), have many advantages.

For starters, this is a digital approach, and produce remarkable realistic sound—sound not subject to tape stretching and/or motor speed variation. And, the chips, being semiconductors, exhibit a far greater life expectancy than magnetic tape. The information is stored inside the chip as part of the physical structure, and is not affected by magnetic fields. Also, their size is smaller than even microcassettes.

Another benefit is the message retention capability of the ICs, especially the EEPROMs (the DRAMS do require battery backup). Once the message is in there, it stays until it is changed—a reasonably simple process. This made the chips very attractive.

However, nothing is perfect, and the ICs suffered from limited recording space. Many of the early systems used EEPROMs for the greeting which were confined to between ten and twenty seconds of speech. But naturally, that would not be sufficient for the caller's message. Thus, the machines retained a magnetic tape cassette for incoming calls.

Even that would change. With advancements in technology, EEPROM chip capacity increased, and this afforded digital recording of both the greeting and the messages. However, as always, it was new, so (here we go again) it was expensive. Today though, a number of completely digital answering machines are available at very competitive prices.

Speaking of today, these gadgets have become almost as versatile as the phones. In fact, they sometimes come with phones. They can also come with memory dialers, call screening, message counters and multiple greetings.

Some use special keypad codes so only the people you really want to talk with can ring the phone. Others will stamp each message with the time and date it arrives. And, of course, many machines handle more than one incoming line.

A few other handy features include voice-actuated recording that conserves tape space, room monitoring which allows you to listen to sounds in the room where the machine is located, toll saver which permits you to hang up when no new messages have been recorded, thus saving long-distance time, and two-way recording that permits both sides of a conversation to be transcribe on tape or digital memory.

Then there is the timer controlled turn on/turn off feature, voice mail capabilities, memo mode that allow you to record quick notes during a call, and the ability to erase and reset your machine from a remote telephone. So, as you can see, the answering machine has come a long way. But, I'll lay odds it hasn't yet completed its entire journey.

Our next device also has a previous background. It is the facsimile machine, and it dates back to before World War II. However, in those early systems, analog techniques were used. The results were acceptable, but slow, so these marvels found a home in areas where photographs needed to be sent over telephone lines. For example, the news wire services and police agencies.

In the 1980s, digital technology again encroached. The result is what most of us think of today as a facsimile, or "fax" machine. What digital brought to this arena was speed and reliability. The speed not only reduced long-distance charges, but also made the machines less susceptible to line noise; hence, more accurate regarding the transmitted information.

**Photo 11-13.** A typical multifunction FAX machine. This model uses thermal paper, but many models are now using standard paper for a more permanent copy.

The first of these digital faxes were like the first answering machines—bulky, and...you guessed it...expensive. And, they were strictly "bare bones"; that is, you fed a document in one side, it was pulled through, reading and sending the text and came out the other side. None of the whistles and bells that we now have.

You had to make the connection through a separate telephone, and the operation was manual only. So, the first addition to many machines was a telephone. In this fashion, you had a "standalone" facsimile system.

Next came automatic connection and digital displays, usually of the liquid-crystal display (LCD) variety. Now, another party could dial your machine and automatically retrieve whatever you had loaded on it. The display would allow you to monitor the numbers you dialed, as well as provide a readout of various system functions. On some units, it also provided a transaction record.

If this sounds a little like a fancy answering machine, that's because it ought to. Faxes and answering machines share a number of features, both technical and operational. But, one of the most important is the ability to detect the ringer signal and seize the line. Once that is accomplished, both can then do their job.

The '80s and '90s also brought numerous individual telephone aids. Among these are amplifiers for the hearing impaired, call restrictors for long-distance blocking, portable DTMF generators for use with "pulse dialing" systems, automatic fax switches that allow a single line to handle both telephone and fax service, and caller ID units that tell you who is calling.

**Photo 11-14.** Two portable DTMF tone generators for accessing dual-tone service.

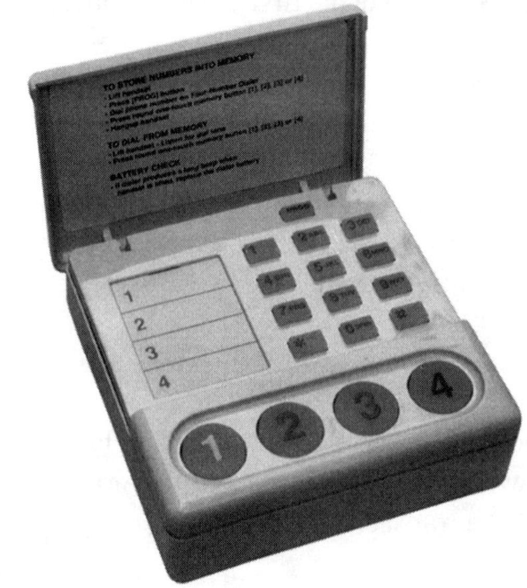

**Photo 11-15.** A handy four-number memory dialer, useful for often-dialed numbers.

Additionally, "hold" buttons, "line in use" indicators, automatic tape recorder switches and "phone flashers" are available. This last one is great for noisy environments, as it detects the ring signal and flashes an incandescent lamp or strobe. That way, you can "see" the phone ringing, even if you can't hear it.

Many distributors also carry a full line of modular jacks, plugs, telephone cable and extensions that permit you to run your own "in house" lines. All of this can save you substantial coins since installation by your local telco is eliminated. That can bring a hefty charge. However, a word of caution. Be sure you know what you're doing, or you could crash your entire phone system—and, make Ma Bell rather mad.

Another area that is well worth looking into, especially if you use your phone in conjunction with a computer, is the "protection" devices. These come in the form of line "spike" shields that absorb transient high voltages, and data guards, or "privacy" protectors. Data guards alleviate the problem of erroneous data transfer caused when someone picks up an extension phone, or call waiting signals your line.

Both of these are important additions to a system, because they either protect you and your equipment from serious shock hazard, or make data transmissions more reliable. And, both come in small, easily concealed packages.

**Photo 11-16.** A handy gadget that protects your phone against telephone line surges. Note: this unit plugs into a wall outlet but the plug spades are plastic. This prevents 120 VAC from being conducted, and the metal ground pin diverts surges to ground.

A last area I want to touch on briefly concerns devices that provide telephone-oriented remote control. This field has garnered neither the attention nor resources devoted to other realms of the industry. But it is, nonetheless, worthy of discussion.

These devices allow you to use your telephone to audibly monitor areas of your home, control appliances, turn lights on and off and so forth. Some even permit automatic dialing of emergency numbers by invalids or others, at the touch of a remote control button—all of which ensures varying degrees of versatility and security.

Unfortunately, telephone equipment retailers do not routinely carry much of this gear. If you are interested in this area, your best bet is probably a home security company. Also, some of the larger department and discount stores handle a limited inventory of items related to this subject.

## CONCLUSION

So, there you have it. In all its glory. Well, part of its glory, anyway, as this is not everything out there. On just about any given day, something new hits the market on the fly.

But hopefully, this chapter has given you a better idea of the multitude of available devices. Naturally, manufacturers will continue to offer novel and/or improved equipment for your benefit and appreciation. And, I doubt any of us will argue too boisterously about that. I know I won't.

For what started out at the Philadelphia Centennial Exposition as a curiosity the public wasn't particularly impressed with, the telephone has traveled a long road. Few devices have emerged from the annals of "modern technology" demonstrating this degree of impact. And, if present-day employment warrants any relevance, the telephone still has a long and healthy voyage ahead of it.

OK! Remember to take that trip to the specialty store. You won't be disappointed. Even if you buy one of everything in the place you won't be disappointed. Just broke. Nah, probably not. This stuff ain't that expensive! Happy shopping!

CHAPTER 12

# A BRIEF HISTORY OF THE TELEPHONE SYSTEM

Throughout this text we have looked at ways to use the telephone system, but how did "Ma Bell" actually come to pass? Most everyone is familiar with the tale of Alexander Graham Bell calling to his assistant, "Mr. Watson, come here, I want you," and Watson hearing this plea over an experimental telephone receiver. The call for help was supposed to have been Bell's reaction to spilling battery acid on his clothes, although, that aspect of the story has been disputed by some. For whatever reason, Bell's utterance did make history, as his words were the first ever heard over the telephone.

This, however, is only a small part of the story. So in this chapter I will present a brief chronicle of the telephone, and the system that supports it. There may be an inaccuracy or two along the way, as history has a habit of becoming embellished. But, for the most part, the following will detail the path that has brought us to where we are today, as accurately as I can convey it.

## THE EARLY YEARS

Alex Bell was born in Edinburgh, Scotland in 1822, and arrived in the Unites States in 1871. He came from a family involved in teaching the deaf to speak, and this, of course, influenced the direction of his life. Ironically, development of the telephone was as much an accident as anything else.

He and his assistant Thomas A. Watson were working on a device that caused metal reeds to vibrate in response to human speech, something Bell called the "harmonic telegraph." The hope was that the hearing impaired could learn to speak correctly by duplicating those reed vibrations.

During one of many experiments, Bell heard a "twang" over the receiver. He rushed to Watson's location and asked what had happened. Watson explained that two spring points had lodged, and he had tweaked the spring to loosen them. This was the sound Bell had heard, and a breakthrough. Not necessarily for the harmonic telegraph, but definitely for the telephone.

With this, the principle of the *telephone* (Greek for "to speak at a distance") had been discovered. Less than two years later, Bell would file for, and be granted, United States patent number 174,465—a patent considered by many to be the most valuable ever endowed, as it initiated an industry that has garnered unimaginable wealth.

In reality, noise, music and telegraph signals had already been transmitted by wire. But voice had not. In that lies the significance of Bell's work. For the first time, his device allowed human speech to be sent over long distances.

While the gravity of this was not immediately appreciated, the advantages the telephone has brought to mankind are hard to challenge. The ability to summon help quickly in emergencies, world leaders being able to discuss problems on a "one-to-one" basis, or just our day-to-day communications all stand as examples of the importance of Bell's invention.

And, we mustn't forget the role telephone lines play in modern-day business and computer systems. Those folks who rely on "E-Mail," on-line services and the internet would be lost without Bell's discovery. More accurately, without Bell's discovery, there wouldn't be E-mail, internet, etc. That's kind of a "chicken-or-the-egg" thing.

Anyway, getting back to the development of the telephone...In June of 1876, Bell and Watson displayed the device at the Centennial Exposition in Philadelphia. Astoundingly, the public wasn't that fascinated.

However, Bell had set up the first phone line between Boston and Summerville, Massachusetts in May of 1876, and by August there were over 800 telephones in use. This did display critically needed interest in his device.

One miscalculation, plaguing original telephone connections, resulted in very poor reception. This involved the use of just a single wire between two phones, allowing a "ground" connection to provide the second conductor. However, the poor quality soon led to the use of two wires, or a "twisted pair," to carry the signals. That action vastly improved the caliber of the signal.

The first commercial telephone exchange went into operation in New Haven, Connecticut. It was set up by George W. Coy on January 28, 1878. With this event, prospects for a credible communications system employing Bell's "newfangled gadget" were realized.

In fact, Alex was so successful in generating interest that in April of 1880, Bell, with the backing of Thomas Sanders and Gardiner G. Hubbard, started the American Bell Telephone Company. By 1887, there were more than 150,000 telephones in the United States. Great Britain had some 26,000, and France 9,000. The age of the telephone had arrived.

It wasn't all smooth sailing, though. Several individuals laid claim to Bell's discovery. For example, a German named Phillip Reis insisted he had invented the telephone first. Then, there was the story of the Pennsylvania farmer claiming he had set up telephones between his house and barn a year before Bell filed for his patent. (I guess he needed to talk to his cows.)

Try as I may, however, I have been unable to secure the farmer's name, or otherwise confirm this yarn. So, for the moment, it must remain just an interesting story. But even if a farmer did invent such a device, he failed to patent it, so the claim is, well, academic.

Finally, after successfully defending himself in over 600 lawsuites, Bell petitioned the U.S. Supreme Court to hear his case. Putting an end to all this, the court found Bell to be the rightful inventor of the device known as the telephone.

Incidentally, on the same day Bell filed for his telephone patent, inventor Elisha Gray filed for a patent on a device he called the "musical telegraph." For all intents and purposes, the musical telegraph was the harmonic telegraph, or the target of Bell and Watson's original research.

In 1885, the American Telephone and Telegraph Company (AT&T) was formed to handle long-distance service for regional telephone companies, or "telcos." Then, in 1900, AT&T absorbed American Bell, forming the telephone giant that's still with us today.

Two additional milestones involve the first switchboard at the Boston office of E.T. Holmes in 1877, and the introduction of the first automatic telephone system in 1891. The second, advertised as the Strowger "girl-less, wait-less telephone," utilized mechanical switches and the first "dial" phones. The slogan was, of course, in reference to the absence of a "live" operator.

## THE SECOND GENERATION

By the turn of the century, the parameters for a telephone system had been more or less established...at least, for awhile. The immediate future held such things as Private Branch Exchanges (PBXs) and some equipment improvements, but for the most part, the basic telephone design would not change radically for the next 60 years, give or take a few years or so.

This did not apply to transmission methods, however. Long-distance service was considered, by those involved, a gold mine, but the 1893 Chicago-to-Boston link seemed to have defined the limit for dependable operation.

Nevertheless, the early twentieth century would bring numerous advancements, ultimately changing that notion. For example, vacuum tubes allowed amplification of the signal along the cable route (land lines). This made it possible to have telephone service wherever lines could be placed.

Additionally, in the late 1930s, tangible work was being done with high-frequency radio waves (50 to 500 megahertz). The early '40s brought the first microwave links (4.3 to 4.8 gigahertz) initially used by the military during World War II.

Microwave moved into the private sector in November, 1947 with the first experimental link between New York and Boston. Although developed by Bell Laboratories, this system was dedicated to television transmission. It wouldn't be until 1950 that a system for telephones was established.

A second major long-distance (long-haul) TV system was rushed to completion for the 1950 World Series. It ran from New York to Chicago and illuminated the problems the vacuum tube was going to contribute to these links. With more than 700 tubes in the system, a failure occurred, on average, every 7 hours.

This caused some inconvenience for the viewers, but overall satisfaction with this new technique seemed to subdue any major complaints. Picture quality was good, and the novelty of being able to watch the World Series "live" in your living room apparently created an air of forgiveness regarding downtime.

Direct long-distance dialing (area codes) made a debut in Engelwood, New Jersey. The year was 1951. The speakerphone arrived in 1955. However, the first speakerphones were expensive, and more popular with businesses (for conference calls) than as a home device.

As stated, the first microwave application for telephones, operating in the 4 gigahertz band, was started in 1950. While only 10 percent of the "circuit miles" (the number of simultaneous calls multiplied by the length of the connection) were carried by microwave, this figure would indeed change. By 1980, 79 percent of all circuit miles were over microwave, and today, regarding long-distance service, these radio links have all but eliminated landline coaxial cable.

While highly efficient, microwaves do have a drawback—that being the frequency at which they operate. In the gigahertz range (billions of hertz), signals are strictly "line of sight" in nature. That is to say, the transmitters and receivers must be able to physically "see" each other.

As mentioned in Chapter 1, this shortcoming means the substations can't be much over 25 miles apart, unless they can be located very high, such as on a mountaintop. Naturally, that restriction requires abundant substations in a long-haul link.

Let's say the link is 1,000 miles long. At 25 miles apart, you will need 40 substations to complete the system. And that is only in one direction. If you want to add additional "sidelinks" (send the signal in more than one direction), these will require supplementary stations at the rate of about 40 per 1,000 miles of signal travel.

Furthermore, this only applies in situations of maximum transmission efficiency. If you encounter disruptive terrain, such as dense forest and/or large cities, you will probably have to add more substations. On the other hand, if mountain ranges cross your path, these can often be used to advantage. The extra altitude can extend the range of the microwave stations.

Well, so much for theory. The purpose here is to illustrate some of the obstacles working microwave systems had to overcome. And, overcome they did.

## THE NEXT STEP

It was now the fifties, and a highly efficient telephone system was in place, both for local and long-distance service. But many problems still existed. Intercontinental communications was done strictly by shortwave radio, a procedure whose roots went back to 1926. And, the same old rotary pulse dials remained in use. Both needed some attention.

Actually, these issues had been addressed in the previous decade. Concepts to improve intercontinental, or transoceanic service came under the category of satellites and undersea cable, while research for a newer, more reliable dialing system had begun in 1941.

Satellites were still the stuff of Buck Rogers, but the idea had been explored by some. In 1945, engineer and science fiction author Arthur C. Clark was the first to suggest the theory, but it would be a few years before technology caught up to him.

Transoceanic, or undersea cable, had garnered some interest, and in 1956 the first such cable was laid between Scotland and Nova Scotia. A third link would later be added between Nova Scotia and Newfoundland. This system works very well, but is a maintenance nightmare. Saltwater reeks havoc on the cable.

An improved dialing scheme began with the "multifrequency key pulsing" (MFKP) approach, but World War II interrupted those efforts. Following the war, in 1948, a small MFKP trial was conducted in Media, Pennsylvania. The test was encouraging, but MFKP used a "damped" oscillation method that didn't produce continuous tones.

This immediately proved unacceptable, and it became clear that continuous tones would be essential for reliable service. However, this first trial did demonstrate that tone dialing was feasible.

With this in mind, engineers began their work. Alas, they needed to overcome a major obstacle involving the tone producing oscillators. The oscillators themselves were not the problem, but making them small enough was. Each standard desk style telephone was going to need seven of them.

So, the next ten years were spent, among other things, designing, testing and perfecting those oscillators. The fruit of this labor would become known as the "dual-tone multifrequency" (DTMF) dialing system. A number of different approaches were tried, but ultimately the 1948 Bell Lab invention of the transistor would be the key.

With this device, researchers could construct very compact oscillators of high reliability. Later, integrated circuits would further enhance both aspects. But until then, transistors permitted engineers to fabricate the necessary equipment to conduct the first DTMF trial.

Two locations were selected to run these 1959 tests; Elgin, Illinois and Hamden, Connecticut. Each site had about 120 active participants, using unique ten-button phones, and the results were deemed promising.

One design feature that proved problematic was the button layout. The ten pushbutton switches were arranged in two rows of five each, and this was awkward to use. So the keypad was rearranged into three rows of three, with a single button as the bottom row. (Later, the pound (#) and asterisk (*) keys would be added, giving a four by three arrangement.) With that change in place, an additional test was conducted.

This trial, in 1960, involved Roanoke, Virginia and Hagerstown, Maryland. The new dial was a vast improvement regarding ease of use and error prevention. AT&T now felt the world was ready for "touch-tone" telephony.

Thus, in 1961 DTMF was introduced as a premium service in Findley, Ohio and Greenburg, Pennsylvania. Within a few years it was available nationwide, and within a decade DTMF was considered "standard" service in many areas. As is said, "the rest is history!" Today, it's hard to find anything but touch-tone phones.

The following year would witness launching of the initial communications satellite, "Telstar I." Other satellites had been put up for the television industry, but this bird was the first to handle telephone service. It was, however, in a "low orbit," thus not stationary. Telstar proved satellites were a viable option, but it swiftly became obvious a geostationary orbit was a must.

Next came two commercial satellites, both in 1965. The first was "Early Bird," succeeded by "Intelsat I." Both operated in the 4- to 14-gigahertz range and admirably managed the industry's workload for almost a decade.

By the mid 1970s, though, service demand was growing rapidly, and several new ventures went into orbit. Leading the way was the 1974 launch of Western Union's "Westar." This was a domestic system, followed by RCA's "Satcom" in 1975 and AT&T's 1976 "Comstar." The later 1970s continued this drill with a vengeance. By 1980 there were in excess of 100 communications satellites orbiting the earth.

The remainder of the '80s and '90s have followed suite, launching both new equipment and replacing satellites that have either left orbit or reached the end of their useful life. By the turn of the millennium, it is estimated that more than 1,000 "celestial servants" will be aloft. Most will provide service for both television and telephones.

## THE LATEST INNOVATION

The 1980s also saw the arrival of a long-awaited mobile communications service. Affectionately called the "cellular telephone," this device is an enormous improvement over previously employed mobile methods.

The original "mobile" telephones used bulky electronic transceivers (transmitter/receiver combinations) that usually had to be installed in the trunk of the vehicle. And, there went half the trunk space. Additionally, this equipment was heavy, fragile, power-hungry and temperamental, while the service was expensive.

Portable applications, such as the "bag phone" and "handheld" we've become accustomed to, were virtually impossible. Size and weight simply precluded that type of equipment.

The principle reason was transmitter output. Unlike the handheld 600-milliwatt or installed 3-watt cellular systems, if these "monsters" were to be practical at all, they required outputs in the 40- to 100-watt range.

Why, you ask? Well, the telephone company didn't have a network of cells (repeater stations) to receive your signal, so range relied on having enough power to reach the regional location—that being the site where the central office transmitting and receiving equipment was situated.

These systems operated in the very high-frequency (VHF) spectrum (often between 147 and 174 megahertz). While the signal displayed reasonably decent carry, only power levels in the aforementioned range proved adequate for clear reception. At that, mobile service was restricted to about 50 miles, even using the highest FCC approved output.

Another problem involved getting on the air. Since very few frequencies were available, it wasn't unusual to experience a half-hour wait for an open line. And, as if that wasn't enough, adjacent cities normally had different frequencies, so if you left your local telco, your mobile phone was useless.

Even so, at the time, the original mobile phone service was highly praised, at least by some. But, the need for a more innovative approach was recognized early on. A number of ideas were explored, but in the end, the cellular concept won out.

The notion of using an overlapping series of repeaters, or cells, first surfaced in the early 1970s. The idea received a very favorable response, but a network of this complexity simply could not be achieved with existing technology.

Aspects such as roaming, channel-scanning and cell hand-off would require dedicated computers, and at that time, machines with the necessary capabilities would have been cost prohibitive. However, over the next ten years all that would change. Intel, Motorola, Advanced Micro Devices and others were hard at work, and low-cost computer equipment was just around the corner.

Hence, by 1980 the first of the cellular systems went into service. As always, it was new, so it was expensive. But that too soon changed. Today, cellular phones have become almost as common as their landline ancestors.

Because you are within a few miles from a cell, at least in most parts of the country, the equipment requires far less power. This, in turn, allows it to be much smaller than its predecessors. And since the system employs "worldwide" compatibility, you can use the phone anywhere you travel. These attributes solved two of the major problems associated with the original mobile telephones.

But that is not the end of the story. Not by a long shot. The 1990s have brought newer, even smaller and more feature-packed phones, as well as "digital" devices. The later apparatus demonstrates a substantial improvement over the more standard "analog" cell phones.

Like anything else digital, information put into the phone (voice, data, etc.) is converted to a "binary stream" of only "0s" and "1s"; that is, the signal is either "on" (1) or "off" (0). This, of course, is a form of code that has to be deciphered, but the effort is worthwhile. The end result, especially reception, is superior in quality.

This improved quality stems from the fact that digital systems force a value to either the "on" or "off" state, thus eliminating the "gray area" (everything in between). And that gray area is the information most likely distorted, or lost altogether.

Another advantage is what might be called *frequency conservation*. The bandwidth (space) needed to transmit just "0s" and "1s" is significantly smaller than required for analog spectrum signals. Consequently, you can squeeze even a greater number of phone conversations onto an existing frequency.

In the end, the signal comes through faster, much clearer, and uses less bandwidth. This is what gives digital cellulars their edge. I wouldn't be surprised to see this technology replace analog phones within the next ten years.

## TELCOS SEE THE LIGHT

The history of telephone service would not be complete without a discussion of fiber optics. Initially a laboratory curiosity, the ability of glass fiber to conduct light, like copper wire conducts electricity, soon caught the attention of many prominent research organizations.

Much of this attention originated in the fields of optics, medicine, physics, photography and of course, telephony. Actually, an understanding of lightwave modulation dated back to Bell himself (the "photophone"), and even before that. But, application of that understanding had never been particularly sensational.

In most cases, researchers attempted to send the light through the atmosphere in what is known as an "open air link." It worked, but even with powerful lamps, optics and reflectors, signal range was nothing short of PIIIITIFUL!

So, when fiber optics emerged in the 1980s, a viable light transfer medium was immediately recognized. Well, almost immediately. After all, this stuff could carry light for miles with very little loss. And, if you modulated that light...*VOILA!*..you had the first optically-based telephone. Needless to say, things weren't quite that simple.

Glass quality and composition, amplification, light sources, cable couplings and receptors were just some of the elements that had to be perfected. However, by the end of the 1980s, a lot of research (and money) helped expedite the first fiber optic telephone service. You remember, "You can hear a pin drop."

A more technical, but nonetheless significant, aspect of this approach involves bandwidth. As the frequency of the "carrier" increases, it can transfer a greater number of simultaneous calls. Since the carrier here is light, the frequency is in the 100-tetrahertz (100 trillion hertz) range.

It is generally accepted that a 14-gigahertz microwave signal can carry 80,000 to 100,000 calls at any given time. So light, at its frequency, should be able to transfer approximately 560,000,000 to 700,000,000 simultaneous calls.

At first, I had difficulty comprehending this concept. If you share my initial confusion, let me present this example in the way of an explanation. Let's suppose that each call requires 1,000 hertz of bandwidth. This means you could carry 1,000 calls in the area between 4 and 5 megahertz (a 1-megahertz span).

However, if the frequency range is between 400 and 500 megahertz, you now have 100 megahertz of space. In this scenario, you can now transfer 100,000 calls. And, when the range is 4 to 5 gigahertz, the span is 1,000 megahertz, or 100,000,000 calls.

As can be seen, with an increase in frequency, the individual band segments become wider and will hold more information. Thus, when you get to light, with segments in the trillions of megahertz, the data capability is almost unlimited.

Naturally, all of this should reduce service expense, if for no other reason than telcos don't have to lay as many lines. Fewer lines should mean less materials, and that ought to translate into diminished initial production cost, less maintenance and lower rates. Or, does it? Rates have seemed to go down in the last decade, but is it due to these reasons, or competition?

With advancements in this field and more cost-effective production (not to mention the ever-increasing cost of copper), fiber optics could, in the not-too-distant future, replace standard coaxial cable. At the very least, optical fiber is going to play an important role, especially in data transfer applications where "quiet" lines are essential.

## CONCLUSION

The preceding was brought to you by your local telephone company. Nah! Not really. But, it has brought you a quick look at the heritage of your local telephone company and telephone service. Hopefully, both are, or will be, helpful.

Unlike computers, integrated circuits, lasers and many other areas of electronics, the telephone is not the product of awesome technology. However, in a sense, this stands as a tribute to those involved in its original development.

In essence, the initial effort was so good it needed little improvement. And today we use a telephone system that has changed only in minor ways since the 1800s; that is, regarding basic principles. I grant you electronics has helped make those principles work more efficiently, but when rendered down, the concept itself has scarcely been modified.

So, next time you use a telephone, you might ponder this chapter. Maybe it will remind you of how the device you hold in your hand came to be. Perhaps this is appropriate. After all, some talented minds brought us to where we are today. And, if you share my interest in the origin of our "creature comforts," hopefully this chapter pleased you.

# CONCLUSION

OK, gang. That "wonderful world of telephones" awaits you! And hopefully this text will help you explore it. Alex Bell's ingenuity led to a vast cosmos of technology that ultimately, and irrevocably, changed our lives. And, most feel that change is for the better. So, pack your provisions and enjoy your journey through this exciting sphere of knowledge. There's a lot to see out there. Don't blink!

I know, I know, that's beginning to get to me too. Enough of the saccharine rhetoric. Really though, the telephone system offers a virtually unlimited arena for us experimenters. And that is one of the primary motivations for this book.

At the outset, I spoke about my personal fascination (sometimes anxiety) with the telephone system and everything associated with it. And over the years, that appreciation has grown stronger; mostly through my own exploration of this field.

However, telephone electronics, like video, can be temperamental. But, once you have gained an understanding of the basic principles, you've gone a long way towards taming the beast. Actually, it's a nice beast, just a little ornery at times.

And, telephony is fun, exciting, rewarding, etc., which makes up for the obstacles. Besides, without obstacles, there wouldn't be any challenge. Right? Right!

Even after you have conquered this field, it still captivates you. The ability to send voice, data and other information vast distances over a pair of wires invites the imagination. For those of us interested in electronics, what greater invitation do we need?

In addition to the "learnin'," this text produces some interesting and useful devices. Hopefully, by now you have tried a few of the projects and found this out for yourself. I especially like the "hands-free" telephone and the telephone lab, but that's just me. I'm sure you will have your own favorites.

So, it's in your hands now, and I have every confidence those are very capable hands. Find the projects you like and "have at" them. I think you will be both surprised and pleased with the results. I'm also sure you will acquire knowledge in the process, and that's not all bad. But, most of all, HAVE FUN with it. That is, and has always been, the most effective incentive for learning!

Shine On!

# ACKNOWLEDGMENTS

I would sincerely like to recognize the following for their assistance in preparing this book. The success and accuracy of this text was greatly enhanced through these sources.

Thanks!
Carl

Compton's Interactive Encyclopedia

World Book Encyclopedia

Telephony: Today and Tomorrow
Dimitris N. Chorafas
Prentice-Hall, Inc. 1984

Encyclopedia of Telecommunications
Robert A. Meyers
Academic Press, Inc. 1989

Gernsback Publications

Communications Circuits Ready-Reference
John Markus
McGraw-Hill Book Company 1987

Ty Fondrin of South Central Bell

# GLOSSARY

**AAA**

| | |
|---|---|
| ADC | Analog-to-digital converter. A device that changes digital signals to the analog state. |
| Alert Tone | The "beep-beep-beep" signal sent to an off-hook telephone that alerts the party to the off-hook condition. |
| Alternating Current | (AC) A voltage source that repeatedly changes polarity. 60-hertz household line is an example. |
| Amperage | Named after the French physicist Andre Marie Ampere. Also called "current," this is the "force" behind a voltage. May be denoted as Amps, Milli- or Microamps. |
| Analog | Any signal that contains continuous information as opposed to "0s" and "1s." Radio, audio and TV are good examples of this form of signal. |
| Analog Meter | The older "moving pointer" style meter that utilizes an inductive movement. |
| Anode | The positive (+) side of a polarity-sensitive component such as a diode. |
| Answering Machine | A device designed to seize the telephone line when the ringer signal appears, give a message, and record an incoming message from the calling party. |
| Astable | A continuous state of oscillation. Self-triggering of the LM555 timer IC is an example. |
| Asynchronous | A communications form in which the data is sent one bit at a time with "start" and "stop" bits to define the message. |

| | |
|---|---|
| Audio | Referring to anything in the human ear sound range of frequencies. |
| Audio Amplifier | An amplifier designed to elevate the level of audio frequencies. |
| Audio Range | The realm of frequencies the human ear can hear, approximately 20 to 20,000 hertz, depending on the ear. |

**BBB**

| | |
|---|---|
| Bandwidth | Refers to the amount of information a "channel" can carry. Expressed in hertz or bits per second, the higher this rating the greater the data capacity. |
| Base | The center of a transistor junction that controls the current flow. |
| Battery Eliminator | A term for any device that provides DC voltage. "Wall Warts" or plug-in supplies are good examples. |
| Battery Voltage | A telco term for the "on-hook" voltage across the "TIP" and "RING" terminals. Usually somewhere around 48 volts DC. |
| Baud | Similar to frequency, baud refers to the number of times switching can occur in a communications system. 300 baud means 300 changes per second. |
| BASIC | "Beginner's All-purpose Symbolic Instruction Code", a "user-friendly" computer language developed by John G. Kemeny and Thomas E. Kurtz of Dartmouth College. |
| Bell, Alexander Graham | The recognized inventor of the telephone. Born in Edinburgh, Scotland (1822), Bell emigrated to the US in 1871, and demonstrated his first working phone in 1876. |
| Bell 103 | A telco MODEM standard that specifies 300 baud, asynchronous communication on two-wire lines. |

| | |
|---|---|
| Binary Coded Decimal | (BCD) A method of arranging long decimal numbers so a computer can process them. The most common example would be the "word" or 8-bit segment of only "1s" and "0s." |
| Binary Stream | The process of sending binary data in either "words" or one bit at a time. |
| Bits Per Second | (bps) Units used to designate the number of "bits" to be sent in one second. Baud rate is one example. |
| Boot | From the expression "pulling oneself up by his bootstraps", it means to initiate the operation of a computer, or computer program. |
| Breadboard | A systematic approach for laying out prototype circuits. The solderless breadboard is a prime example. |
| Buffer | For hardware, a circuit that boosts the output of a device. With software, a memory area that temporarily holds information. |
| Busy Signal | A Central Office imposed signal that lets you know the number you have called is in use. |

**CCC**

| | |
|---|---|
| Cable | In this context, cable is the wire(s) that carry telephone calls. |
| Calibrate | Any process that aligns an operation or measurement with the expected value — adjusting the reference potentiometer on a digital voltmeter, for example. |
| Call Restrictor | A device that prevents a specified type of phone call from being made. |
| Capacitor | A common electronic component that has the ability to store a charge for a short time. Also used to prevent DC flow. |

| | |
|---|---|
| Cathode | The negative side of a polarity-sensitive component, such as a diode. |
| CCITT | An acronym for Consultative Committee for International Telephone and Telegraph, it is an international group that sets standards for telephone communications. |
| CD | (Carrier Detect) Pin 1 of the 9-pin serial port that handles the "line present," or carrier input signal. |
| Cell | In this context, a substation that receives from and transmits to wireless telephones while connecting them to the Central Office. |
| Cellular | Term used to designate a "mobile phone" system that employs a network of cells to generate full telephone service. |
| Central Switching Office | Abbreviated CO, the first destination of a phone call. From here, the various "on-line" tones and signals are produced and routing of the call is accomplished. |
| Channel | A specified frequency spread within which information can be contained and transmitted as a single unit. Also refers to the outputs of a DTMF signal controller. |
| Chip | Slang for "integrated circuit." |
| Circuit Mile | A telco term for the number of simultaneous calls multiplied by the connection distance. For example, 25 calls traveling 100 miles would be 2,500 circuit miles. |
| Clamp | In this context, means to suppress a signal or voltage. Dual zener diodes in the line interface are an example of a "ringer signal" clamp. |
| Clark, Arthur C. | An engineer and science fiction writer who in 1945 proposed the possibility of satellite communications. |

| | |
|---|---|
| Coaxial Cable | A transmission line that utilizes an outer shield to prevent loss and noise. Very popular in the telephone industry. |
| Code | With computers, this refers to the numerical language they understand. Synonymous with computer program. |
| Coil | An electronic component comprised of wired wound around a center core. This device produces a magnetic field and is often called an electromagnet. |
| Cold Joint | A poorly executed solder joint that is characterized by a dull "flaky" appearance, and poor conductivity. |
| Collector | The side of a transistor junction that receives the current flow. |
| Communications | The process of conveying information from one point to another. |
| Communications Satellite | Complex electronic systems stationed above the earth (outer space) to send and receive data. This scheme provides very long range signal transfer useful for both telephones and television. |
| Complementary Metal-Oxide Semiconductor | (CMOS) A type of integrated circuit (IC) architecture that provides low power consumption over a wide range of voltages. Frequency response is dependent on operational voltage with a limit of about 10 megahertz. |
| Controller | Any device that regulates operation of another device or function. |
| Converter | A circuit that changes a signal from one electronic state to another. Analog to digital is a good example. |
| Coupling Capacitor | Employing a capacitor in between two stages to provide smooth interaction of the stages. Also, usually performs DC voltage blocking. |
| Coy, George W. | Operated the first telephone exchange in 1878. |

Crystal — An electronic component made from quartz that resonates when electricity is passed through it. Used primarily for frequency control.

CS — (clear to send) Pin 8 of the 9-pin serial port that inputs a signal from the host computer that it is ready to receive data.

**DDD**

D-Sub Connector — A universal interconnection system that employs pins and sockets to connect peripherals to a computer. These come in 9, 15, 25, 36, 50 and other pin arrangements.

DAC — (Digital to Analog Converter) Device takes a digital string of "1s" and "0s" and changes it into an analog signal.

DC Block — Any method that prevents direct current flow. Capacitors are the most commonly used device for this purpose.

DCD — (Data Carrier Detect) The computer side of the DC (data carrier) signal line in a 9-pin serial connection.

Decoder — Any device that deciphers binary-coded decimal signals. The DTMF receivers are examples of this operation.

Dedicated Line — With telephone service, this is a specialized line that handles specific tasks. These can be high-speed MODEM lines or super quiet data transfer links, as well as others.

Demultiplexer — A circuit that separates input signals into individual outputs respective of the input signal's code.

Dial Tone — A dual tone pair of 350 and 440 hertz that the CO imposes on the line when the phone is first taken "off-hook." This signal will disappear after the first digit is dialed.

| | |
|---|---|
| Digital | Referring to forcing the value of data to either a zero (0) or a one (1), as opposed to infinitely many values between zero and one. This process allows microprocessors to evaluate and process said data. |
| Digital Multimeter | (DMM) An electronic testing device that incorporates several different measurement tools. Common DMMs can measure voltage, amperage, resistance, with some adding frequency and capacitance to the equation. |
| Diode | An electronic component that rectifies alternating current (AC) to direct current (DC). Also used for radio signal detection. |
| Direct Current | (DC) A voltage source with a fixed polarity. Power supplies and batteries are two primary examples. |
| Display | A device used to visually provide information. LED and LCD are two technologies commonly used for this purpose. |
| DPDT | (Double Pole Double Throw) Designates a switch that can "throw" two signals in one of two directions. |
| DPST | (Double Pole Single Throw) Designates a switch that "throws" two signals in only one direction. In effect, a double "on-off" switch. |
| DS | (Data Set ready) Pin 6 of the 9-pin serial port that tells the computer the information is ready to send. |
| Dual-Tone Multifrequency | Abbreviated DTMF, this is the dual tone scheme used by modern telephone systems to dial a desired phone number. Each digit it represented by one "high" audio tone and one "low" audio tone. |
| "Dumb" Terminal | Refers to a communications system that is manually-operated (you know, by a warm body) as opposed to "data control" signals automatically regulating the system. |

| | |
|---|---|
| Duplex Adapter | A dual or sometimes multi-output modular jack that expands the capability of the wall telephone line access to more than one device. |
| Dynamic Random Access Memory (DRAM) | A type of semiconductor memory that is fast but requires a constant source of power to retain its information. |

**EEE**

| | |
|---|---|
| E-Mail | "Electronic" mail, a widely-used form of communications that takes advantage of computers and computer networks instead of your local post office. |
| Electret Microphone | A type of microphone that uses a capacitor as the "pickup" element. By detecting the change in capacitance, due to vibration, the mic is capable of perceiving sound with high sensitivity. |
| Electrically Erasable and Programmable Read-Only Memory | (EEPROM) A semiconductor memory that is able to hold data without power for extended periods of time. The data can be both electrically recorded and erased as opposed to EPROM chips that have to be erased with UV light. |
| Electronic Ringer | An electronic circuit that requests you answer the phone as opposed to the old "bell & clangor" announcement of days gone by. |
| Emitter | The side of a transistor junction that supplies the current flow. |
| Encoder | Any device that generates coded data to be passed on to another device. The DTMF generators used for dialing are a good example. |
| Entrance Bridge | (Also called "terminal bridge") An electrical block that accommodates the incoming telephone line. |

**FFF**

| | |
|---|---|
| Farad | The basic unit for capacitance, named after Michael Faraday. |
| FAX | (facsimile) In our context, a machine that can transmit and receive documents and photos via the telephone system. |
| Fiber Optics | The science of bending light by reflecting it off the interior of a glass fiber. Used extensively in the telephone industry. |
| Fiber Optic Cable | Cables made up of strands or "fibers" of super high-quality glass. Since glass is technically a liquid, these strands are quite flexible, making the cables highly pragmatic. |
| Field Effect Transistor | (FET) Specialized semiconductor characterized by very high input resistance. |
| Field Phone | A telephone system that supplies its own electrical support thus making it completely portable as well as independent from established telcos. Originally developed for the military battlefield communications. |
| Filter | An electronic circuit that restricts certain frequencies. Also used to remove noise from power supplies and audio systems. |
| Frequency | Designated in "hertz," it represents the number of "cycles" that occur in one second. Can be Kilohertz, Megahertz, Gigahertz and so forth. |
| Frequency Counter | A specialized "event" counter that utilizes a timing period or "window." By counting the number of cycles that occur during that window, the frequency of the source can be determined and digitally displayed with LED/LCD 7-segment arrays. |
| Full Duplex | Term refers to the ability of a communications system to both send and receive simultaneously. A standard "land-line" telephone is an example. |

| | |
|---|---|
| Gain | Term used to gauge the amount of signal increase or amplification a circuit or chip achieves. |
| Gray, Elisha | Man who eventually patented the device Bell was working on when he invented the telephone. Called the "musical telegraph," ironically, Gray filed the patent the same day Bell filed for his discovery. |
| Gremlin | A perhaps mythical creature that can't be seen or heard but lives in your home and creates all manner of trouble. Some folks are skeptical of their existence, saying the trouble is really your own fault, but I Don't Know! |
| Ground | A common "rail" or "leg" in a circuit, usually negative in polarity, that provides collective feedback for the circuit. Often, a substantial number of circuit components will have at least one lead connected to ground. |

**HHH**

| | |
|---|---|
| HAM | No, not smoked pork, but a frequently-used term for amateur radio operators. |
| Hands Free | Term applied to devices that leave your hands free for other duties. The "headset" style telephone is such a device that allows you to talk on the phone and still work with your hands. |
| Harmonic Telegraph | The device Bell and Watson were working on when the telephone was invented. It would have helped the hearing impaired learn to speak by teaching them to reproduce vibrations in metal reeds located within the device. |
| Henry | The basic unit of inductance, named after American Joseph Henry. |
| Hertz | Honoring German electrical pioneer Heinrich Hertz, it is the basic unit for frequency, or cycles per second. |

| | |
|---|---|
| Holmes, E.T. | Had the first switchboard in his business in 1877. |
| Hook Switch | The switch in a telephone that determines whether the phone is "on-hook" or "off-hook". Can be as simple as a SPST for one side of the line, or as complicated as a DPDT that connects/disconnects both sides. |
| Hubbard, Gardiner G. | Helped start the American Bell Telephone Company in 1880. |

**III/JJJ**

| | |
|---|---|
| IBM | International Business Machines. One of the world's largest computer development companies. |
| Impedance | Gauged in ohms, the inductive resistance to the flow of AC. Usually refers to speaker coils or antennas. |
| Inductor | Another name for a coil, or device in which magnetic fields are created by current flow. |
| Input | The signal put into a circuit, or the place where the signal is put into said circuit. |
| Integrated Circuit | (IC) Any one of a huge number of active devices in which an entire circuit is etched onto/into a silicon wafer. The result usually allows complex equipment to be built with a minimum of components. It also greatly enhances the reliability of the finished device. |
| Interface | Any arrangement that allows one circuit to connect to another. Can be as simple as just a cable, or as complicated as an entire third circuit. The matching transformer based system is one example of a "telephone line" interface. |
| Inverter | Also known as the "inverting buffer," this logic "gate" changes its input to the opposite value. Thus, a high input results in a low output. |

| | |
|---|---|
| Jack | Any one of numerous receptors that accepts a matching plug. For example, an earphone jack on a radio. |

**KKK/LLL**

| | |
|---|---|
| Keypad | A type of keyboard that uses fewer keys than the standard computer keyboard. The telephone dialing pad for example. |
| Laser | (Light Amplification by Stimulated Emission of Radiation) An optical device that produces a coherent, collimated beam of light. Usually of primarily one color, this light can travel long distances without much beam spread or intensity loss. |
| Laser Diode | Specialized semiconductor diode that produces true laser light. These have become extremely popular as the light source for fiber optic communications. The light is easily modulated and bears all the properties of more standard laser illumination. |
| Latching Relay | A type of mechanical relay that stays in position after the magnetic coil is deactivated. Another signal is used to release it. |
| Light Emitting Diode | (LED) A specialized diode who's junction glows when an electrical current is passed through it. The junction is often a combination of arsenic and gallium, and can produce light of just about every color. |
| Line Level | A semiamplified signal level between the lower microphone output and the higher power amp output. Normally, if a device will drive a small speaker, it is considered "line level." |
| Line of Sight | Refers, in this context, to the characteristic of ultra and super high-frequency radio waves carrying only as far as can be seen. This is somewhat of a disadvantage with "overland" systems, but a big advantage with satellite systems, as the signals will travel great distances using very little power. |
| Linear Regulator | One of the most common ICs used to regulate voltages. While not especially efficient, are highly accurate and easy to use. |

Link — Used here to describe any connection between two communications devices. Satellite link, for example.

Liquid Crystal Display — (LCD) Display arrays that use chemicals known as "liquid crystals." These chemicals have the property of changing color when their molecules are "twisted" by electricity and become visible in whatever pattern they are arranged.

Load — The device being controlled by a "controller" circuit. A relay, motor, LED and SCR are a few examples.

Local Area Network — (LAN) A computerized communications system set up within a "local area" such as an office or business. Used heavily to share peripherals like printers.

Local Loop — The network existing between you and your local Central Office. It does not include any long-distance service or systems.

Long Haul — A term for telecommunications over any long distance. Usually refers to calls made outside your local area.

**MMM**

Matching Transformer — An "isolation" transformer used in telephone equipment. Usually, the primary and secondary are both 600 ohms which complies with telco requirements for line quality and seizure.

Metal-Oxide Varistor — (MOV) A suppression device that is used to protect telephone equipment against line "spikes." Such surges are often induced by lightning.

Metal-Oxide Semiconductor Field-Effect Transistor — (MOSFET) A specialized transistor characterized by very high input resistance and a dual "gate" arrangement.

Microphone — Any of many differently designed electronic sensors that detect audio sound.

| | |
|---|---|
| Microwave Link | Radio repeater stations that utilize the "microwave" or giga hertz radio spectrum. Very efficient transmissions, but strictly line of sight in nature. |
| Mobile Phone | Today usually refers to cellular service, but can also mean any radio-based telephone system. The term originated with the older VHF car style transceiver telephones. |
| MODEM | (MOdulator-DEModulator) The device that made the Internet possible. By converting digital signals to analog and back again, the MODEM allows computers to use the analog telephone system to move data. |
| Modular Cable | A length of telephone cable with a modular plug on each end. Used to connect phones/devices to the system. |
| Modular Jack/Plug | A specialized receptor system designed for telephone service. Characterized by a molded plug that only fits the companion jacks. |
| Modulation | The process of attaching information to any "carrier" style signal. This can be by wire or radio waves. |
| Multi-Frequency Key Pulsing | (MFKP) The first "tone dialing" scheme set up and tested by the telephone company in the 1950s. Displayed several problems that eventually led to DTMF. |
| Multifunction | In this context, refers to telephones that possess more than one function. Combination answering machine-telephone systems are one example. |
| Multiline | Refers to a telephone capable of handling more than one individual telephone line, or number. |
| Multimeter | Any measuring device with either a digital or analog meter that can manage more than one measuring function. Volt-Ohm Meter (VOM) is an example. |

**NNN**

| | |
|---|---|
| Neon Lamp | An "ionization" lamp where electricity applied to electrodes in an envelope of neon gas causes the neon to glow. |
| Network | In this context refers to a system of interconnected communications devices, such as telephones or computers. This ain't NBC, or one of the others. |
| Nibble | (nybble) In computer lingo, this is a four-bit word, or one half a standard eight-bit "byte." You don't take a bite, you take a nibble. Get it? |
| Noninverting | Any device, chip or system that doesn't change the state of the input signal, such as a "buffer" gate or one input of an operational amplifier. |
| NOR Gate | A logic gate in which the action is the reverse of the NAND gate. With NAND all outputs are high except when both inputs are high. With NOR, all outputs are low until both inputs swing low. |
| "NOT" Gate | Applies to any logic gate that results in an output opposite of the input. NAND, NOR and inverters are all prime examples. |
| Off-Hook Voltage | Usually in the 4- to 7-volt range, is the voltage present across the Tip and Ring when the phone is taken off-hook. |
| Off-Hook | The physical act of lifting the telephone "receiver" off the hook switch. Hey, I thought you might want to know! |
| Ohm | The basic unit of resistance, named for German mathematician George Simon Ohm. |
| On-Hook | OK, let's watch the remarks. This is the physical action of placing the telephone receiver on the hook switch. Again, I just thought you would want to know about this. |

| | |
|---|---|
| Operational Amplifier | Any one of a whole slew of ICs that performs amplification duties. Some are for just audio frequencies while others include radio frequencies. All like to see a split power source, but most will work without it. |
| Optoisolator | A "line" isolation device that uses LEDs and photosensors to control a circuit from one side of the device to the other. For example, when the LED lights, it causes a phototransistor to allow current to flow, thus energizing a circuit. |
| Oscillator | A device that produces a repeating cycle pattern at a specific frequency. The LM555 IC is an example. |
| Output | The signal emitted by devices such as amplifiers, oscillators, power supplies, etc. |

**PPP**

| | |
|---|---|
| Passive | In this context, any action that is purely for information as opposed to being an active part of the operation. Dial tones and "on-off" LEDs are two illustrations. |
| Perf-Board | Circuit construction material that has a uniform pattern of holes for component installation. Due to standard lead spacing, the holes are often 0.1 inches apart. Some boards also have copper soldering pads on one side for easier parts placement. |
| Phase-Locked Loop | (PLL) An electronic circuit that controls frequency by locking the signal into a specific "loop" pattern. The LM567 tone detector is an example. |
| Photophone | One of Bell's many experiments, this device sent voice over a light beam. While functional, it was plagued with problems due mostly to the type of light sources available at the time. |
| Piezo Element | Utilizing the principle that crystal material vibrates when an electric current is imposed upon it, these "buzzers" produce high-volume sound, and are used extensively in electronics. |

| | |
|---|---|
| Plug | The "male" half of any specific receptor system. The power plug on your computer for instance. |
| Point-To-Point Wiring | The process of connecting electronic components with individual wires as opposed to the copper "traces" of a printed-circuit board. |
| Polarity | With DC-oriented electronics, this refers to the positive (+) and negative (-) poles of the power source or components involved. |
| Positive | The plus (+) pole of a battery, DC supply, electrolytic capacitor or other polarity-sensitive item. |
| Potentiometer | A variable form of resistor that allows a change in resistance. Volume controls are a prime example. |
| Power Supply | Any device that provides the working potential for anything electric or electronic. Can be a battery, transformer-based supply, generator and so forth. |
| Power Transistor | A specialized transistor designed to handle high voltages, current or both. |
| Printed-Circuit Board | (PCB) Fiber or epoxy-based circuit board that has copper traces etched on one or both sides for conduction of the circuit's electrical power. For very stable and sturdy circuit construction. |
| Prototype | The first, or experimental, version of a design or idea. With electronics, this often begins on a solderless style breadboard. |
| Pulse | Refers to a short signal, usually regarding digital electronic circuits. |

**QQQ**

| | |
|---|---|
| Quartz | A form of silicon dioxide (SiO2), it is the preferred choice of material for crystals and piezo elements. |

Quiet Line — A telephone line which is highly filtered to keep extraneous "noise" to a minimum. Important in "data" transfer systems to minimize false signals.

**RRR**

R/C Tank — (Tuning) A method of tuning an oscillator in which a resistor (R) and capacitor (C) are employed in a parallel arrangement to set the frequency.

RD — (Receive Data) Pin 2 of the 9-pin serial port that allows data to be input to a communications device.

Receiver — In this context refers to DTMF decoders that receive tone pairs and convert them to a binary-coded decimal (BCD) output.

Rectifier Bridge — A full-wave rectifier circuit contained in a single package. This device changes alternating current (AC) to direct current (DC).

Regulator — Any device/system that holds the circuit voltage to a specific level. May be either positive or negative.

Reis, Phillip — The German who insisted he invented the telephone before Bell. His argument never held up, however.

Relay — An inductive-based device that either opens or closes switch contacts when current is passed through its coil.

Resistor — Probably the most commonly used of all electronic components, it prevents the flow of electricity. Resistors are important in balancing power consumption.

Ring Generator — A device that produces 90 to 110 volts AC at 20 hertz. Used to activate the "ringer" in a telephone.

Ring Detector — A circuit/device that detects the presence of a ring generator signal. Can be an optoisolator-based system as well as other schemes.

| | |
|---|---|
| RING | The red wire of the four-wire telephone cable. The name refers to the outer "ring" of a phone plug used with the original manual switchboards. |
| Ringer Equivalence Number | (REN) The power rating the telephone company assigns to each device used on its system. With most home telephone lines, the telco allots a rating of 5 or 6 for the total REN, however if your requirement exceeds that rating, if requested, the phone company will usually increase it. Adding the RENs from each device on your line will reveal your requirements. |
| Ringer | Look out! This is a toughy! This is a device inside your telephone that "rings" when someone is calling you. I warned you! |
| Ringer Signal | The 90- to 110-volt AC 20-hertz signal sent by the Central Office to activate your telephone's ringer. Another complex piece of work. |
| RJ11 | The manufacturing designation given to the "modular plugs/ jacks" that provide telephone service to your home, business or elsewhere. |
| Rogue | An unscrupulous person with low morals and ethics who might invade your life. Sort of like your Uncle Oscar. I include this subject only for your protection and well-being. |
| Rotary Dial | Ah Yes! I remember it well! That old round plate with holes that you had to stick your finger, or a pencil, or something into and rotate to dial whatever telephone number it was you wanted to dial. How's that for long-winded? The result was a series of "pulses" that told the CO what number you were trying to reach. |
| RS-232 | An asynchronous communications standard recommended by the Electronic Industries Association (EIA). This is your computer's serial port. If it's not, then you need a new computer. |

## SSS

| | |
|---|---|
| Sanders, Thomas | In 1880 helped start the American Bell Telephone Company which was later absorbed by American Telephone and Telegraph (AT&T). |
| SD | (transmit data) Pin 3 of the 9-pin serial port that returns data to the computer. |
| Semiconductor | Any one of many electronic devices that is made from silicon substrate. These include diodes, transistor and, of course, integrated circuits. |
| Serial Port | The computer's communications port that allows data transfer via the serial or "one bit at a time" method. Usually a 9-pin arrangement, it can also be 25 pins in size with the major difference being power connections and additional control lines. |
| Short Haul | As opposed to the long haul over great distances, this is usually local connections over short distance. Although, in very small communities this might include connection to another small community. |
| Side Link | Usually refers to alternate microwave lines off a main microwave substation. Allows added service from the primary link. |
| Sidetone | A passive phone company feature. By diverting some of the microphone sound signal to the telephone speaker, you can hear yourself, and that assures you the system in functioning. |
| Sidetone-Balance Network | In the older nonelectronic telephones, the sidetone-balance network was a box inside containing capacitors, resistors and inductors that provided the "sidetone" effect. |
| Silicon Control Rectifier | (SCR) A semiconductor device, similar to the transistor, that latches after activation. It then takes a second signal to release the SCR. Useful for applications where you want the device to stay activated. |

| | |
|---|---|
| Smart Terminal | An asynchronous communications device that responds to control signals for "automatic" operation. |
| Software | A method of coding information so a computer can understand it. Software is written in "languages," such as BASIC, Fortran or C++, that facilitate this process. |
| Solder Bridge | The dreaded condition of solder crossing between two separate pads or traces on a circuit board. This, of course, causes an unwanted connection and trouble. |
| SPDT | (Single Pole Double Throw) A switch that sends a single signal in one of two directions. |
| Speaker | An inductively-based electronic device that turns electric impulses into sound by vibrating a fiber or metallic cone. |
| Speaker Phone | Nothing more than a standard telephone with a second microphone and speaker built into the case. This allows the user to hang up the receiver and still carry on a conversation. Usually these systems are sensitive enough to pick up most sound in a normal size room or office. |
| Split Power Supply | Provides both positive and negative voltages. Useful with analog-to-digital converters and operational amplifiers. |
| SPST | (Single Pole Single Throw) A switch that throws one signal in one direction. The simple "on-off" switch is one such device. |
| Substation | A major telephone link separate and away from the Central Switching Office. Microwave long-distance repeater stations are excellent examples. |

**TTT**

| | |
|---|---|
| Telco | (TELephone COmpany) That's all folks. There ain't no more! |

| | |
|---|---|
| Telephone Terminal Block | (Also known as the "entrance bridge") The point at which the telephone line enters the building. Usually a plastic box with snap lid and binding posts for the red, green, yellow and black line leads. |
| Telephone | I was unable to locate any suitable definition, and/or description of this device. Thus, I'll have to leave this one up to you. Nah, just kidding! You know what a telephone is! (Right?) |
| Telephone System | Collectively, all the equipment, devices, cables, fiber optics and anything else that comprises all the telephone companies. |
| TIP | The green lead of the four-wire telephone line. Its name comes from the "tip" of the phone plug used with the original manual switchboards. |
| Tone | An audio frequency signal used for dial tones, alert tones, busy signals and, of course, DTMF dialing tone pairs. |
| Tone Decoder | A circuit, often IC, that converts the DTMF tone pairs to 4-bit binary-coded decimal (BCD) nibbles used to activate other circuits. |
| Tone Generator | A circuit, often IC, that produces the DTMF dual tones needed for "touch-tone" dialing. Sometimes called a tone encoder. |
| Tone Pair | In the DTMF scheme, two separate audio tones are assigned to each individual digit. These are known as "tone pairs". |
| "Touch Tone" | Bell copyrighted name for dual-tone multifrequency (DTMF) tone dialing. |
| Transformer | Inductive device that transfers energy from one coil to another. *Power* and *matching* are two varieties of this component. |
| Transient Suppression | Removal of unwanted noise, signals or spikes on a line. Capacitors and varistors are two commonly employed devices for this purpose. |

| | |
|---|---|
| Transistor | A semiconductor junction that can provide both switching and amplification functions. The primary component of integrated circuits, and the "solid state" replacement of the vacuum tube. |
| Transistor-to-Transistor Logic | A form of integrated circuit logic (TTL) architecture that provides high-frequency response (around 20 megahertz), but has a narrow voltage range (4.75 to 5.25 volts) that must be strictly observed. Otherwise you'll end up with "green fried silicon" or something equally distressing. |
| Transmitter | A circuit that sends a carrier signal and/or information via radio waves, light or cable. In this context, microwave and satellite systems are example of telcos using transmitters. |
| Twisted Pair | No! This is not some new hard rock group out of El Segundo. This refers to the pair of wires that carry telephone signals. They are often literally "twisted" to minimize noise and interference. |
| Two-State | In this context refers to the alternating tones of an electronic telephone ringer. |

**UUU/VVV**

| | |
|---|---|
| Ultra High Frequency | (UHF) I'll may get an argument here, but this is the radio frequency spectrum between ABOUT 400 megahertz to ABOUT 800 megahertz. If you have a different spectrum, so be it. |
| Undersea Cable | Also know as "transoceanic," this was the first alternative to the noisy radio transmission intercontinental telephone connections. Coaxial cable was actually laid on the ocean floor to carry the calls, but maintenance was a nightmare. Satellites have all but replaced this form of service. |
| Vacuum Tube | The original amplification, rectifying and switching component, tubes are constructed of metal plates, grids and filaments housed in an "evacuated" glass envelope. They served us well for many years, but were fragile, hot and somewhat unreliable. Thus, they have been mostly replaced by semiconductors. |

| | |
|---|---|
| Very High Frequency | (VHF) The radio frequency spectrum from ABOUT 50 megahertz to ABOUT 400 megahertz. As with UHF, I expect some disagreement. |
| Volt | The basic unit of electromotive force, named after Italian physicist Count Alessandro Volta. |
| Voltage Spike | A sudden surge of high voltage that can be damaging to equipment and/or components. Lightning and power company glitches are common causes. |
| Voltage Divider | A resistor network that divides an input voltage, by a ratio determined from the resistor values, to a lower voltage. |
| Voltage Doubler | A capacitor/diode arrangement that doubles, and rectifies, an input alternating current (AC) voltage. |

## WWW—ZZZ

| | |
|---|---|
| Watson, Thomas A. | Alexander Bell's assistant who was instrumental in the development of the telephone. |
| Watt | Named after Scottish scientist James Watt, this is the "brute" behind the voltage. It actually started as a formula to define "power" as 1 amp flowing through 1 volt. |
| Wide Area Telephone Service | (WATS) A form of long-distance telephone service where tolls are based on a "flat-rate" charge. Very handy for business purposes, as I'm sure everyone is aware. |
| X-Y Matrix | A keypad scheme where a keystroke connects a "column" to a "row." This approach is used for standard telephone keypads. |
| Zener Diode | A specialized diode that is designed to regulate voltage levels. This is done by "reverse biasing" the diode. However, that requires special diode construction, as most diodes will "burn up" under this condition. |

# Index

## G



## H



## I



## K



## L



## M



## N



## O



## P



## R



## S

# T



# U - Z

Your Technology Connection to the Future!

A Bell Atlantic Company

Howard W. Sams

# *Now You Can Visit Howard W. Sams & Company On-Line: http://www.hwsams.com*

*Gain Easy Access to:*

- **The PROMPT® Publications catalog, for information on our *Latest Book Releases*.**
- **The PHOTOFACT® Annual Index.**
- **Information on Howard W. Sams' Latest Products.**
- ***AND MORE!***

PROMPT®
PUBLICATIONS

PROMPT®
PUBLICATIONS

## Electronic Projects for the 21st Century

*by John Iovine*

If you are an electronics hobbyist with an interest n science, or are fascinated by the technologies of he future, you'll find *Electronic Projects for the 21st Century* a welcome addition to your electronics li-rary. It's filled with nearly two dozen fun and use-ul electronics projects designed to let you use and xperiment with the latest innovations in science and echnology. This book contains the expert, hands-on uidance and detailed instructions you need to per-orm experiments that involve genetics, lasers, ho-ography, Kirlian photography, alternative energy ources and more. You will obtain all the informa-ion necessary to create the following: biofeedback/ ie detector device, ELF monitor, Geiger counter, MHD generator, expansion cloud chamber, air pol-ution monitor, laser power supply for holography, inhole camera, synthetic fuel from coal, and much nore.

Projects
256 pages • paperback • 7-3/8 x 9-1/4"
ISBN: 0-7906-1103-1 • Sams: 61103
$24.95

## RadioScience Observing Volume 1

*by Joseph Carr*

Among the hottest topics right now are those related to radio: radio astronomy, amateur radio, propagation studies, whistler and spheric hunting, searching for solar flares using Very Low Frequency (VLF) radio and related subjects. Author Joseph Carr lists all of these under the term "radioscience observing" — a term he has coined to cover the entire field.

In this book you will find chapters on all of these topics and more. The main focus of the book is for the amateur scientist who has a special interest in radio. It is also designed to appeal to amateur radio enthusiasts, shortwave listeners, scanner band receiver owners and other radio hobbyists.

RadioScience Observing also comes with a CD-ROM containing numerous examples of radio frequencies so you can learn to identify them. It also contains detailed information about the sun, planets and other planetary bodies.

Communications Technology
336 pages • paperback • 7-3/8 x 9-1/4"
ISBN: 0-7906-1127-9 • Sams: 61127
$34.95

**CALL 1-800-428-7267 TODAY FOR THE NAME OF YOUR NEAREST PROMPT PUBLICATIONS DISTRIBUTOR**

Prices subject to change.

PROMPT®
PUBLICATIONS

## Desktop Digital Video

*by Ron Grebler*

*Desktop Digital Video* is for those people who have a good understanding of personal computers and want to learn how video (and digital video) fits into the bigger picture. This book will introduce you to the essentials of video engineering, and to the intricacies and intimacies of digital technology. It examines the hardware involved, then explores the variety of different software applications and how to utilize them practically. Best of all, *Desktop Digital Video* will guide you through the development of your own customized digital video system. Topics covered include the video signal, digital video theory, digital video editing programs, hardware, digital video software and much more.

Video Technology
225 pages • paperback • 7-3/8 x 9-1/4"
ISBN: 0-7906-1095-7 • Sams: 61095
$34.95

## The Video Hacker's Handbook

*by Carl Bergquist*

Geared toward electronic hobbyists and technicians interested in experimenting with the multiple facets of video technology, *The Video Hacker's Handbook* features projects never seen before in book form. Video theory and project information is presented in a practical and easy-to-understand fashion, allowing you to not only learn how video technology came to be so important in today's world, but also how to incorporate this knowledge into projects of your own design. In addition to the hands-on construction projects, the text covers existing video devices useful in this area of technology plus a little history surrounding television and video relay systems.

Video Technology
336 pages • paperback • 7-3/8 x 9-1/4"
ISBN: 0-7906-1126-0 • Sams: 61126
$29.95

**CALL 1-800-428-7267 TODAY FOR THE NAME OF YOUR NEAREST PROMPT PUBLICATIONS DISTRIBUTOR**

Prices subject to change.

PROMPT® PUBLICATIONS

## The Complete RF Technician's Handbook, *Second Edition*

*Cotter W. Sayre*

This is THE handbook for the RF or wireless communications beginner, student, experienced technician or ham radio operator, furnishing valuable information on the fundamental and advanced concepts of RF wireless communications.

Circuits found in the majority of modern RF devices are covered, along with current test equipment and their applications. Troubleshooting down to the component level is heavily stressed. RF voltage and power amplifiers, classes of operation, configurations, biasing, coupling techniques, frequency limitations, noise and distortion are all included, as well as LC, crystal and RC oscillators. RF modulation and detection methods are explained in detail – AM, FM, PM and SSB – and associated circuits, such as AGC, squelch, selective calling, frequency multipliers, speech processing, mixers, power supplies, AFC, multiplexing, cellular and microwave technologies.

Communications Technology
366 pages • paperback • 8-1/2 x 11"
ISBN: 0-7906-1147-3 • Sams: 61147
$34.95

## The Right Antenna *Second Edition*

*Alvis J. Evans*

Television, FM, CB, cellular phone, satellite and shortwave signals are available in the air to anyone, but it takes a properly selected and installed antenna to make them useful. With easy-to-understand text and clearly illustrated examples, *The Right Antenna* will give you the confidence to select and install the antenna that meets your needs.

*The Right Antenna* explains how antennas work and breaks them into TV and FM for discussion. A separate chapter is devoted to interference, and antennas used by hams for antenna band operation. This new second edition also includes chapters on DSS, other satellite antennas, and TV and amplifier antennas.

Also included: How Antennas Work, Selection of Antennas for Specific Needs, Installation of Antennas, Fringe Area and MATV Antennas, How to Identify and Eliminate TV Interference.

Communications Technology
120 pages • paperback • 6 x 9"
ISBN: 0-7906-1152-X • Sams: 61152
$24.95

**CALL 1-800-428-7267 TODAY FOR THE NAME OF YOUR NEAREST PROMPT PUBLICATIONS DISTRIBUTOR**

Prices subject to change.

PROMPT® PUBLICATIONS

## Speakers For Your Home and Automobile, *Second Edition*

*Gordon McComb, Alvis Evans, Eric Evans*

The cleanest CD sound or the clearest FM signal re useless without a fine speaker system. *Speakers or Your Home and Automobile* will show you the ows and whys of building quality speaker systems or home or auto installation. With easy-to-understand instructions and clearly-illustrated examples, his book is a must-have for anyone interested in mproving their sound systems.

The comprehensive coverage includes:

peakers, Enclosures, and Finishing Touches
onstruction Techniques
Viring Speakers
utomotive Sound Systems and Installation
Iome Theater Applications

Audio
192 pages • paperback • 6 x 9"
ISBN: 0-7906-1119-8 • Sams: 61119
$24.95

## Complete Guide to Audio

*John J. Adams*

*Complete Guide to Audio* was written for the consumer who wants to know more about sound systems. With comprehensive, simple explanations, it answers questions you may have asked salespeople in the past but were unable to get answers for.

In addition, this book explains some common problems that you may run into while setting up your home entertainment center. The information in the book will help you make successful purchasing decisions and demystify the jungle of wires and connections that come with your audio system.

Topics include: Audio Basics, Sound, Stereophonics, Home Theater, Amplifiers and Preamplifiers, Receivers and Surround-Sound, Cassette and CD Decks, DVD, MiniDisc and Phonographs, Speakers, Headphones and Microphones, Computer Sound, Brands and Choices, Hookups and Accessories.

Audio
163 pages • paperback • 7-1/4 x 9-3/8"
ISBN: 0-7906-1128-7 • Sams: 61128
$29.95

**CALL 1-800-428-7267 TODAY FOR THE NAME OF YOUR NEAREST PROMPT PUBLICATIONS DISTRIBUTOR**

Prices subject to change.

PROMPT®
PUBLICATIONS

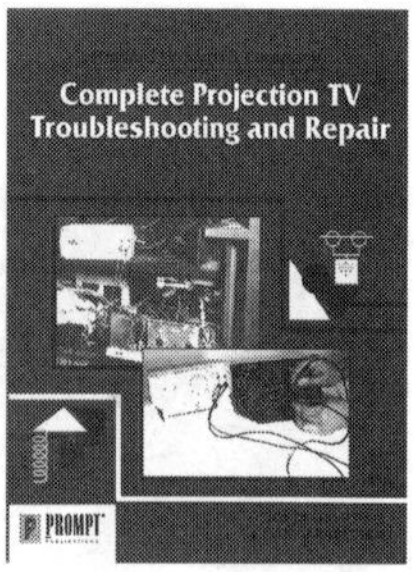


## Complete Projection TV Troubleshooting and Repair

*Joe Desposito and Kevin Garabedian*

*Complete Projection TV Troubleshooting & Repair* is just the tool every professional technician, enthusiastic hobbyist or do-it-yourselfer needs to successfully repair projection TVs.

This book covers everything from the basics of projection circuits, tools and test equipment, TV types and measurements, to the finely detailed repair techniques. It also shows how to troubleshoot tuner/demodulator circuits, audio and video circuits, horizontal and vertical circuits, high-voltage circuits, microprocessor circuits, power supplies, CRTs, projection TV controls and remote control circuits.

This book also contains a case study for the Mitsubishi Model VS-458RS – a step-by-step repair with photographs and detailed schematics.

Troubleshooting and Repair
184 pages • paperback • 8-1/2 x 11"
ISBN: 0-7906-1134-1 • Sams: 61134
$34.95

## 1998 Computer Monitor Troubleshooting Tips

The information in this book will save hours of troubleshooting time, the cost of searching for schematics or tips, and the irritation of slow repairs.

*1998 Computer Monitor Tips* includes over 2000 troubleshooting and repair tips listed by manufacturer and model number, featuring major names such as Apple, Acer, Compaq, CTX, Dell, Gateway, Goldstar, IBM, NEC, Packard Bell, Sony, Viewsonic, Zenith and many more. Repair procedures for the KD1700 series monitor, including FREE schematics and parts list is also included.

It also contains FREE FCC number identification software to cross-reference model information to the original manufacturer and contact information, as well as an overview of how VGA monitors work, an overview of monitor EEPROM repair, an overview of ESR in-circuit capacitor testers and international monitor manufacturer contact listings.

Troubleshooting and Repair
252 pages • paperback • 8-1/2 x 11"
ISBN: 0-7906-1160-0 • Sams: 61160
$49.95

**CALL 1-800-428-7267 TODAY FOR THE NAME OF YOUR NEAREST PROMPT PUBLICATIONS DISTRIBUTOR**

Prices subject to change.

PROMPT®
PUBLICATIONS

## IC Design Projects

*Stephen Kamichik*

*IC Design Projects* discusses some of the most opular and practical ICs, and offers you some rojects in which you can learn to create useful and nteresting devices with these ICs.

Once you have read through this book and com-leted some of its projects, you will have a real, work-ng knowledge of ICs, enabling you to design and uild you own projects!

Topics include: how power supplies operate, in-egrated circuit voltage regulators, TTL logic, CMOS ogic, how operational amplifiers work, how phase-ocked loops work, and more!

Projects include: battery charger, bipolar power upply, capacitance meter, stereo preamplifier, func-ion generator, DC motor control, automatic light imer, darkroom timer, LM567 tone decoder IC, elec-ronic organ, and more!

Projects
261 pages • paperback • 7-3/8 x 9-1/4"
ISBN: 0-7906-1135-X • Sams 61135
$24.95

## Build Your Own Test Equipment

*Carl J. Bergquist*

Test equipment is among the most important tools that can be used by electronics hobbyists and profes-sionals. Building your own test equipment can carry you a long way toward understanding electronics in general, as well as allowing you to customize the equipment to your actual needs.

*Build Your Own Test Equipment* contains infor-mation on how to build several pragmatic testing devices. Each and every device is designed to be highly practical and space conscious, using com-monly-available components.

Projects include: Prototype Lab, Multi-Output Power Supply, Signal Generator and Tester, Logic Probe, Transistor Tester, IC Tester, Portable Digital Capacitance Meter, Four-Digit Counter, Digital Multimeter, Digital Function Generator, Eight-Digit Frequency Counter, Solid-State Oscilloscope, and more.

Professional Reference
267 pages • paperback • 7-3/8 x 9-1/4"
ISBN: 0-7906-1130-9 • Sams: 61130
$29.95

**CALL 1-800-428-7267 TODAY FOR THE NAME OF YOUR NEAREST PROMPT PUBLICATIONS DISTRIBUTOR**

Prices subject to change.

PROMPT®
PUBLICATIONS

## Electronic Circuit Guidebook, Volume 1: Sensors

*Joseph Carr*

Many texts discuss instrumentation and computer nterfacing. A topic usually lacking in these texts, ıowever; is covered in *Electronic Circuit Guidebook, /olume 1: Sensors* - namely, the analog interface.

In this volume you will find information you need ıbout typical sensors, along with much more infornation about analog sensor circuitry. Amplifier cir:uits are well covered, along with differential am›lifiers, analog signal processing circuits, and more.

Topics covered include: Noise, Sensors and Intruments, Instrument Design, Resistive Sensors, Ǝlectro-Optical Sensors, Magnetic Sensors, Inter'aces, Building the Analog Subsystem, Analog Amps, nverting and Noninverting Amps, Differential Λmps, Analog Signal Processing Circuits, Removng 60 Hz Noise, Shield and Grounding Methods, Λ/D-D/A Converters, Sensor Resolution Improvenent Techniques, DC Supply Circuits, AND MORE.

Electronics Technology
339 pages • paperback • 7-3/8 x 9-1/4"
ISBN: 0-7906-1098-1 • Sams: 61098
$29.95

## Electronic Circuit Guidebook, Volume 2: IC Timers

*Joseph Carr*

Part I of this book demonstrates the theory of how various timers work. This is done by way of an introduction to resistor-capacitor circuits, and in-depth chapters on TTL and CMOS digital IC timers, the LM-555 and other ICs, operational amplifier timer circuits, retriggerable timers and long duration timers. The simplified equations and detailed graphics are perfect for both technicians and hobbyists.

Part II presents a variety of different circuits and projects, some standalone, and others for incorporation into other circuits. Examples included: touch-plate trigger, missing pulse detector, analog audio frequency meter, one-second timer/flasher, relay and optoisolator drivers, two-phase digital clock, 100kHz crystal calibrator and many more. *Electronic Circuit Guidebook, Volume 2: IC Timers* will teach you enough to be able to rework and modify the circuits covered here, and also design a few of your own.

Electronics Technology
239 pages • paperback • 7-3/8 x 9-1/4"
ISBN: 0-7906-1106-6 • Sams: 61106
$29.95

**CALL 1-800-428-7267 TODAY FOR THE NAME OF YOUR NEAREST PROMPT PUBLICATIONS DISTRIBUTOR**

Prices subject to change.

PROMPT®
PUBLICATIONS

## Electronic Circuit Guidebook, Volume 3: Op Amps

*Joseph Carr*

Third in the series, *Electronic Circuit Guidebook, ⁄olume 3: Op Amps* is designed to give some insight nto how practical linear IC amplifiers work in acual, real-life circuits. Because of their widespread ›opularity, operational amplifiers figure heavily in his book, though other types of amplifiers are not ›verlooked. Volume 3 allows you to design and conigure your own circuits, and is intended to be a pracical workshop aid.

Topics covered: Linear IC Amps, The Ideal Op ⁄mp, Inverting Follower Op Amps, Noninverting ⁄ollower Op Amps, Ideal vs. Non-Ideal Op Amps, )C Differential Op Amps, Instrumentation Amps, solation Amps, Operational Transconduct-ance ⁄mps, Current Difference Amps, Solid-State Audio ⁄ircuits, Wideband and Video Linear IC Amps, Acive Analog Filters, Waveform Generators and Timer ⁄ircuits, Instrumentation and Other Circuits.

Electronics Technology
273 pages • paperback • 7-3/8 x 9-1/4"
ISBN: 0-7906-1131-7 • Sams: 61131
$29.95

## Electronic Circuit Guidebook, Volume 4: Electro-Optics

*Joseph Carr*

Volume 4 is mostly about E-O sensors – those transducers that convert light waves into a proportional voltage, current or resistance. The coverage of the sensors is wide enough to allow you to understand the physics behind the theory of the device, and also the circuits used to make these sensors useful. This book examines the photoelectric effect, photoconductivity, photovoltaics, and PN junction photodiodes and phototransistors. Also examined is the operation of lenses, mirrors, prisms, and other optical elements keyed to light physics.

Volume 4 also examines other devices, including optoisolators, optical fibers, X-ray sensors, TV sensors, and additional optical instruments such as telescopes and scientific instruments. The circuits for using sensors in E-O instruments, and for interfacing them with computers, are also discussed in depth to allow you to apply them to practical situations.

Electronics Technology
405 pages • paperback • 7-3/8 x 9-1/4"
ISBN: 0-7906-1132-5 • Sams: 61132
$29.95

**CALL 1-800-428-7267 TODAY FOR THE NAME OF YOUR NEAREST PROMPT PUBLICATIONS DISTRIBUTOR**

Prices subject to change.

# PROMPT® PUBLICATIONS

## TV Video Systems for the Hobbyist and Technician

*L.W. Pena & Brent A. Pena*

Covering standard hard-wired cable, large-dish satellite systems, wireless cable, and DSS (Digital Satellite Systems), *TV Video Systems for The Hobbyist & Technician* explains the different systems, how they are installed, their advantages and disadvantages, and how to troubleshoot problems.

Presented in easy-to-understand text and illustrations, it covers installation instructions, home options, apartment options, detecting and repairing problems, and more, guiding you through your TV video project to a successful conclusion.

*TV Video Systems for the Hobbyist & Technician* covers the following: The Different TV Video Systems, Contracting for TV Video Services, Typical System Components, Apartment Installations, Home Installations, How To Recognize and Overcome Problems, Video Installations–The Right Way, Adding Accessories, and Troubleshooting and Repairing.

Video Technology
124 pages • paperback • 6 x 9"
ISBN: 0-7906-1082-5 • Sams: 61082
$19.95

## Complete Guide to Video

*John Adams*

*Complete Guide to Video* explains video technology systems in an easy-to-understand language. It outlines the common components of modern audio/video equipment and gives details and features of the newest gadgets. This book was designed to help you with your shopping choices and to answer questions about video technology that the average salesperson may not know. An invaluable reference guide, this book will help you set up the right components for your surroundings and will give you the information you need to get the most out of your video equipment.

Topics include: Home Video Basics, Sights, Sounds & Signals: All About TVs, Antennas, Satellite Dishes & Cable, VCRs, LaserDisc Players & DVD Players, Camcorders, Video & Audio Editing, Game & Internet Consoles, Purchasing Help & Recommendations, Features, and Troubleshooting.

Video Technology
336 pages • paperback • 7-3/8 x 9-1/4"
ISBN: 0-7906-1123-6 • Sams: 61123
$29.95

**CALL 1-800-428-7267 TODAY FOR THE NAME OF YOUR NEAREST PROMPT PUBLICATIONS DISTRIBUTOR**

Prices subject to change.